# ANY FOOL CAN BE A
# DAIRY FARMER

Books by James Robertson

*Published by Farming Press*

Any Fool Can Be a Dairy Farmer
Any Fool Can Be a Pig Farmer

*Published by Pelham*

Any Fool Can Be a Villager
Any Fool Can Be a Yokel
Any Fool Can Be a Countryman
Any Fool Can Be a Countrylover
Any Fool Can Keep a Secret
Any Fool Can See a Vision

# ANY FOOL CAN BE A DAIRY FARMER

## JAMES ROBERTSON

Drawings by
CHARLES GORE

Farming Press

*First published 1980*
*Reprinted 1981, 1982, 1989*

British Library Cataloguing in Publication Data

Robertson, James, *1945–*
  Any fool can be a dairy farmer
  1. Devon. Dairy farming. Personal
  observations
  I. Title
  636.2'142

  ISBN 0-85236-195-5 Paperback
  ISBN 0-85236-110-6 Cased

Farming Press, 4 Friars Courtyard, 30–32 Princes Street,
Ipswich IP1 1RJ, United Kingdom

North American Distributor:
Diamond Farm Enterprises, Box 537,
Alexandria Bay, NY13607

Printed and bound in Great Britain by
Biddles Limited, Guildford and King's Lynn

# Preface

The first batch of cows to come to the farm clattered out of the cattle lorry. They scattered out over the field and proceeded to graze on the rich Devon grass. I, as proud new owner, climbed over the gate and strolled amongst them. One of them lifted her head as I approached and came towards me. She craned her neck and blew great gouts of warm, sweet-smelling breath over me. She rasped her tongue along my outstretched hand and rubbed her head on my leg as I leaned over to scratch her back, a deep affection for her welling within me. She turned her flank towards me and looked back, her great plum-coloured eyes surrounded by their fringe of impossibly long eyelashes. Then, with a precise and delicate gesture, she lifted her back hoof and kicked me neatly in the balls. In that brief encounter, she encapsulated the tortuous love-hate relationship that I enjoyed with a herd of cows over five years.

# Chapter One

I entered the stone-flagged farmhouse kitchen, leant against the Aga and thoughtfully sipped my tea. 'Hmm,' I said. My wife looked up in alarm from the baby. She was doggedly trying to stuff some particularly revolting looking baby food down its throat.

'Oh no,' she said, 'I know that "Hmm."'

'Well, I was just thinking . . . '

'That's what I was afraid of.'

' . . . that we've done advertising.'

'Yes.' She put down the spoon and started to concentrate. The baby surreptitiously spat its mouthful on to the floor. The dog casually rose from its basket by the Aga and cleared it up.

'And we've done living in London.'

'Yes, and you've had a bash at law.'

'That's not really true. You couldn't call it much of a bash. I didn't get as far as qualifying.' The dog was now sitting alertly under the pine kitchen table waiting for more droppings. The baby fixed it with a cold eye and regurgitated something unspeakable. The dog quivered with anticipation and its tail started to thrash against the floor with excitement.

'No, I suppose not,' she replied.

'And now we've done pig farming.'

'You don't think we ought to stick at it a bit longer?' The dog stretched a hopeful nose out towards its titbit and received a rattle in the chops.

'I don't think there is very much point. We've been doing it for three years and still haven't made any money. We'd have to spend a fortune on the buildings to make this place viable.'

'I suppose that's one excuse.' My wife could sometimes have a regrettable view of my capabilities. 'Still,' she went on, 'it's certainly been worth while, hasn't it?'

'Sure. It's been very interesting. But I was wondering what we ought to do next.' The dog was now flat on its stomach and crawling beneath the chair legs to attack its goal from another direction. My wife dug the spoon into the bowl and dangled it thoughtfully in front of the baby who screwed up her face. The dog froze.

'Why not become a brain surgeon? Prime Minister? Undertaker? Bookie?' She was beginning to get into the spirit of things.

'They all sound a bit routine and citified.'

'Well, how about throwing pots, or growing tomatoes, or doing the self-sufficiency bit?'

'It's possible I suppose. But there's something a bit earnest and humourless about that sort of thing.' My behind was becoming uncomfortably warm against the stove. I moved over to the sofa in the corner of the room and sat down. That kitchen was large enough to act as sitting room and spare bedroom besides.

'Anyway they all sound like hard work.' I suppose that after a few years of marriage a wife does get to know her partner.

'Too right. You know, I still fancy a go at dairy farming. I haven't quite got agriculture out of my system yet.'

'I know, dear, but we've been through all that. You need a dirty great farm to keep cows and this place has only four acres. To buy a dirty great farm, you need a dirty great pile of money. And going bust in pigs is not the best way of going about making one.' The baby had become tired of the vaguely proffered spoon and thrust out an arm at it. The contents described a parabola in the air and the dog frantically tried to wriggle out from under the chair to intercept them. They landed stickily on the back of his neck. The baby crowed in glee.

'Hmm.'

'Why don't you become an antique dealer? That seems to support lots of odd people.'

'You're not accusing me of being odd, I hope.'

'Oh no. Not really. You must admit that it was a bit odd to want to become a pig farmer in the first place.'

'Nonsense. It was just as much your idea in the first place.' I was damned if I would let her shovel all the responsibility on me. The dog had retired in a huff to his basket and the cat joined him and started to wash down the back of his neck. 'I suppose,' I went on, 'we could always go back to Scotland and I could get a steady job or something.'

'What an appalling idea.'

'I've got to do something that will bring in a living.'

The conversation really died there and I went down to the estate agent the following day to put the farm on the market. However, the gods that guard the irresponsible did not abandon us. A farm tenancy came up.

The tenancy of a farm is a jewel beyond rubies. Every one that comes up attracts hordes of applicants, qualified to their eyeballs with their grandmothers already wrapped for sale on the off-chance that it would help. Against competition like that, what hope had I? No qualifications, no agricultural training or experience at all, save a brief and unsuccessful dabble in pigs. I scarcely knew how big an acre was and even that had just been made redundant by some mysterious thing called a hectare. I had one slight advantage over everybody else. The owner was my father-in-law.

He had bought the farm in Devon some ten years previously while he had still been working in London. He lived there at weekends and enjoyed being a gentleman farmer. After retiring, he no longer needed a tax loss which showed little prospect of turning into a profit and so he decided to sell. Then I struck. To be more truthful, my neighbour struck. We had already resolved to return to commerce and, initially, took only limited notice of father-in-law's decision to sell. My neighbour was a dairy farmer. Why, he suggested, did I not offer to become a tenant? No, he did not consider that my sublime ignorance was an insurmountable handicap. He could teach me how to farm before we moved down there.

My prospective landlord agreed to consider the notion and asked for a tender which would also demonstrate how I would farm the land. My neighbour was delighted to do it for me. I sent it off without understanding a word that was written. Agreement came by return. We could move in in three months.

Right. So I was now an embryo dairy farmer. Things from now on had to be taken in a logical progression. I had to sell the pig farm in order to raise some capital with which to buy cows. This was already under way, so that should be no problem. The other thing to do was to find out how to dairy farm. This my neighbour was to take care of and so I signed up as a farm worker in order to learn the trade.

I turned up in pristine new green gumboots, bright eyed, bushy tailed and filled with enthusiasm and determination to suck his brain dry of information and experience. He pressed a fork into my hand, pointed at a shed and said 'Muck it out.' So I meekly mucked. The shed had been inhabited by calves over the preceding winter. Calves have to be kept clean and healthy and so straw bales are frequently chucked at them to give them something comfy to lie on. Over the course of a winter there is a build-up — in this case a build-up of about four feet — of tightly packed reeking straw, laced with several hundred gallons of dung and urine.

The shed was rectangular with a door at one end. Twenty per cent of the muck could be tossed straight out of the door. The other eighty per cent could either be carried to the door and then ejected or, more excitingly, shot precisely through the missing pane of a window that had rusted closed. It is one of the prime rules of farming that all windows on all farms should have missing panes.

Mucking out a calf shed is an art form that is rarely appreciated by outsiders. You drive your fork into the mass of straw and try to lever out a wedge. A true forkful will, you will find, weigh about 65 lb. Lever at it and you will snap your fork. You soon learn to judge your wedges. Anything over 35 lb and the amount of energy expended in moving it would be better employed in handling two lighter loads. You achieve your precise 35 lb, balance it carefully and sway slightly. Then you uncoil in a spasm of effort directed at the missing window pane. With practice, you should be on target five times out of six. The sixth will hit the wall and stick there rather like a plateful of spaghetti before slithering tiredly down.

It took me four days to empty this shed. I was joined by a visitor to the farm, who would park himself on a shooting stick in a part of the shed already cleared and smoke endless cigarettes whilst treating himself to a monologue on the decay of moral values since the last war. At the end of the fourth day, I returned to my employer for further instructions, looking hopefully at his cows. He thought for a moment, handed me back my fork and pointed at the next door shed. 'Muck it out.' I mucked my way through three more sheds and could now earn a living by lecturing on moral decay. And then release — I was promoted. A field was indicated.

'Roll it', quoth the sage. So I rolled.

Rolling a field is one of the labours of Sysiphus. It was progress on mucking out because the work effort was reduced; but the tedium was beyond belief. I was given precise instructions. Collect a tractor. Couple up the roller. Proceed to the indicated field and roll.

I found the tractor. First start the brute: ignition on, turn key. Nothing happened. I plodded back to enquire why. 'You first have to have the gears in neutral before the engine will turn over.' Eureka, I thought, until the engine had repeatedly turned over and still nothing had happened. I plodded back to enquire why. 'Did you use the heater plug?' 'What's a heater plug?' 'It's a thing that preheats the cylinders so that the engine will start.' I preheated like mad and nothing happened. I plodded back to enquire why. 'Did you switch on the fuel?' No I had not switched on the fuel. I switched on the fuel and the battery was flat. With infinite patience, my employer pull-started the tractor.

Now to couple up the roller. I backed the tractor up to the roller and heaved the connecting bar out of the mud and tried to slip in the tow pin. The two holes did not marry. I heaved at the roller but it weighed a couple of tons. So I re-manoeuvred the tractor. The one-inch diameter holes still did not meet. It was about then that I almost broke down and sobbed, resolving not to become a dairy farmer, but grit won through and, eighth time, I married the holes, slipped in the pin and went to roll a meadow.

For efficient rolling of meadows, it is necessary to travel no faster than walking pace. If it is a good meadow, you appear to achieve precisely nothing. If it is a bad meadow, you get bogged down in a corner. My first meadow was a good meadow. It was also very large and extremely cold. I went up and down it for an hour or two. Then I changed and went down and up it. A few hours of that and I threw caution to the winds and went round and round it. It took a couple of days to roll that field and then I rolled another four or five and then I complained to the management. Well, no, he was damned if he would let me near his cows. After all his living depended on them, but I could build him a block wall.

Building a block wall was something I could do, having spent many happy hours at it as a pig farmer so that the pigs could knock it down again. This particular block wall was thoroughly absorbing because, apart from being the largest that I had

11

tackled, I was asked to build it beside a rat's hole which gave an added dimension to the proceedings. The rat fancied a go at the feed bags near by and I fancied a go at the rat. I would pretend to be totally absorbed in the wall and the rat would tiptoe out from his hole. I would then spin round and hurl a dollop of mortar at him; he would give me a dirty look and retire back to his hole for an hour or so before he would creep out again. Neither the rat nor I ever tired of this game which lasted a good fortnight and then I finished the wall and returned to the boss for instructions.

I was now progressing because I was put onto fertilising and then sowing and yet more rolling. Time was moving on a bit and so my neighbour transferred me to the one job that all pig farmers are experts in — cleaning and fumigating buildings. As consolation, he lent me his set of college notes and through these I began to learn all about the diseases of the cow, or some of them, as the handwriting became a little obscure during some of the more exciting passages when the emotion of the moment during note-taking had caused a lack of concentration on legibility. Agricultural colleges are nothing if not thorough and I learned about tractors and sugar beet, how to replace pulleys and the insulation values of common materials. Like much theory, most of it turned out to be grossly irrelevant and what was relevant usually turned out to be wrong, but from the point of view of building up the confidence of an aspiring dairy farmer, the notes were beyond rubies.

I never did manage to get my hands on a cow. I watched the boss milking for fifteen minutes, but he threw me out as he said it made the cows nervous. I would have persevered but my student days were up as the three months was over and it was time that I took up with our destiny.

At the beginning of April, we gathered our goods and chattels and made our move. Like all moves, it was a major operation, made even more so in our case as we were bringing down all sorts of agricultural bits and pieces that had been used for pigs in the hope that they would have some application for cattle. My wife and family went ahead to blaze the trail while I superintended the loading of a gargantuan pantechnicon and drove down in an ancient Hillman Imp. My passengers were the dog and two cats, both sitting rather sulkily in bird cages.

The cats were very much farm rather than domestic, used to the wild blue yonder and keeping as far from human company as possible. Both took an extremely dim view of car travel. Senior cat waited until we were bowling down the motorway at 45 mph before, with a sound like ripping plaster, it defecated. The dog was nearest to the scene and turned white and started to whimper. A whiff crept between the seats, presaging a wave of a stench so appalling that I cut in front of a lorry onto the hard shoulder and baled hurriedly out of the car with the dog in hot pursuit. A blue-lamped Range Rover cruised up behind.

'Having a spot of trouble, sir?'

'Yes, the cat has, er, shat.'

'You mean that your car has not broken down and you pulled off the motorway on to the hard shoulder merely because of the cat?'

'Yes.'

'That is not a good enough reason, sir.'

'Right, officer. You stick your head into that car and then have the courage to tell me that it is not a good reason.'

The policeman looked uncertain but sent his mate to investigate. His mate managed to open the door wearing that patient, pitying expression that is used by the constabulary when faced with the more simple sections of the public. He bent forwards and nearly lost his cap as he jerked his head back and stepped warily away.

'Yeah, well,' he said, taking a deep breath, 'I think we ought to leave the gentleman to sort out his little problem.' Senior policeman looked a little annoyed and stepped towards the car. His mate caught his eye and shook his head warningly.

'All right, sir; but get going as soon as you can and try not to block the hard shoulder except in emergencies.' They stepped well round the Hillman and roared off.

It took ten minutes to fumigate the car and clear up the cat. We went on our way again and it was at least fifteen minutes before a triumphant explosion from junior cat heralded a repeat performance. The dog spent the rest of the trip half way out of the window, and I nearly caught pneumonia.

The farm, seventy acres in all, was well off the main tourist routes of Devon, near a village described in the gazetteer as being high and remote. From a 'B' road, six miles from the nearest

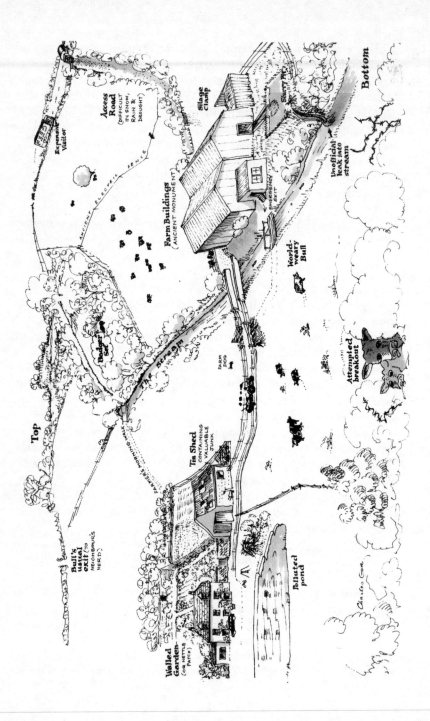

market town, we plunged into a maze of narrow lanes that wound their way between high hedge banks. The farm itself lay at the end of a half-mile track that plunged down a hill. The surface of the track had been worn down over the centuries by the constant passage of the hooves of generations of cattle and, above it, the hedges almost met creating a dark tunnel with its sides studded by campion and primroses and criss-crossed by the tracks taken by deer, badgers, foxes and rabbits.

Down in the valley, a small stream ran under the lane which jumped across it on a narrow, concrete bridge. Beside the stream were the farm buildings fronted by a mellow, stone and cob cowshed roofed with warm clay tiles. Up the side of the stream ran a tangled wilderness of marshy woodland, studded with ferns, mosses and lichens. Here lived the badgers, while water voles plopped in and out of the pools and buzzards and ravens nested in the higher trees.

Up from the stream, on the opposite side, was the farmhouse, its front smothered in Virginia creeper. Built about a hundred years earlier, it had replaced the original thached Devon longhouse, with its cowshed at one end and the living quarters at the other, which had been destroyed by fire. From the house we could look out across the stream to the fields which rose on the far side of the valley towards the public road. A fringe of tall beeches defined the farm's boundary on the horizon, topped by a tumultuous cawing rookery. To the left of the farm lane which could be seen winding down the hillside, the small irregularly-shaped fields stretched towards a solitary dead oak that stood sentinel against the skyline.

Beyond the farmhouse, the lane ran through another farm and out again to another public road which ran up to the village.

This, then, was the asset which we were supposed to exploit and the first step was to decide how to go about it. We needed cows, food for cows, facilities for the cows to live in and facilities to milk the cows. We were not at all sure how to go about any of these steps. That was a feeling that I was beginning to know quite well. Believing as I do, that the only way to operate is by jumping in at the deep end and finding out whether it will be sink or swim, the only uncertain moment occurs in the middle of the jump, before the waters overwhelm you and you are too concerned with the business of survival to think about anything else. It is not so much a feeling of uncertainty as one of flat panic as you cast around desperately for that which will initiate you into the new way of life. One large question mark lay in the fact that my predecessor who, by all accounts, had been an excellent stockman had not produced a profit from the farm. There was I, a rank amateur, expecting to succeed where the expert had failed.

The first problem was sheep. There were sheep all over the farm. It was April and they had clipped the grass on the farm down to the level of a bowling green. Even I knew enough to understand that the cows would need that grass, and it might already be too late to save enough to feed the cattle over the coming winter. We had to evict them fast. Sheep farmers who have their stock safely grazing on another's land in the middle of the growing season know that they are on to a good thing and make themselves very scarce when the parish tomtoms start to beat out the message that the new tenant wants his land cleared. An ultimatum was delivered after a couple of fruitless weeks. Either the sheep would be removed within three days, or the field gates would be opened and the sheep would be driven into outer darkness. It worked, and in dealing with sheep farmers it seems to be the only way that does work.

The fields were then empty and hopefully growing grass for the cows that were to come. So attention was turned to the buildings. We were really quite well off once we had evicted another farmer's scores of rat-like little bullocks that lurked in the darker corners. There was an enormous corrugated iron barn supported on a dozen tree trunks, into which we were supposed to put silage or hay. There was a modern shed split into cubicles for winter accommodation for the stock and a block containing milking parlour, dairy and collecting yard where the cows would stand and queue for entry.

The one lesson that my boss in Wales had drummed into my skull was that a milking parlour should be as efficient and as easily run as possible. Milking is a twice-a-day chore and he was strongly of the opinion that all milking should be done through a herringbone parlour. A herringbone consists of a hole in the ground in which the operator stands, with the cows on the level alongside with their udders at a convenient milkable height. He who has to bend down to milk, ran the theory, is so knackered that he cannot do anything else but rest and recover until it is time for the next milking.

I summoned my first expert from the Ministry of Agriculture, which is in business to dole out grants to indigent farmers such as myself so that we can improve our working conditions. In those days, there was still some sense in grants. Now the Ministry gives grants in order to encourage milk production, while the EEC gives grants in order to bribe farmers out of milk production. The Ministry man came to look at the shed where my pit was to be dug.

'There's no room for a herringbone.'

'What do you mean — no room for a herringbone? It's a dirty great rectangular shed into which a dirty great rectangular pit would fit very nicely.'

'But,' he replied, 'the cows would have to enter the building on the long side of the rectangle.'

'So?'

'Cows don't turn corners and they would have to turn a corner once they had entered the parlour so that they could go into the stalls.'

'Don't be daft. Are you going to refuse to give me a grant because you think that cows can't turn corners?'

'Something like that. You could always demolish the building and start again.' Aha. This was familiar territory. I have found that if you apply for official money for anything, it invariably ends up by costing twice as much as if you had not. I was damned if I would become involved in that particular game.

I had an idea. 'Suppose the cows didn't have to turn a corner. Then it would be OK?'

'Yes.'

'Right then, the cows won't turn corners.'

'But they will.'

'No they won't. Go back to your office and assume that you never came out here. I shall pretend that I will put another door in so that they come in straight and you can approve the plan.'

Bless his heart. He agreed to this and I got my money. The cows turned the corner like corkscrews with no bother at all and everybody was happy.

I hired a man to dig the hole and lay out the parlour. He was highly experienced because he had dug his first hole for a neighbour a few weeks earlier.

'I want my hole to be thirty-six inches deep,' I told him.

'Thirty-six, boss? That's far too deep. They should be no more than twenty-six.'

'Should they indeed? But I am quite tall and I pay the bills so I would suggest you make it thirty-six inches deep.'

'You're the boss, Boss.' He dug his hole and it was twenty-six inches deep.

'I wanted my hole to be thirty-six inches deep.'

'No, I couldn't have done that. It would have been far too deep.'

'But that's far too shallow.'

'Nonsense, Boss. You can always put in a few duck boards to change the level and give you another inch or two.' While I was trying to work that one out, he tugged his forelock and slipped smartly away. He was the cause of much anguish in the future because his shallow pit meant that various highly sensitive portions of my anatomy projected above the edge and were vulnerable to twitching hooves.

With progress being made on the building, it seemed to be about time to think about buying cows. Apart from the fact that most of them were black and white and all should have udders, I really knew very little about them and such ignorance boded ill for effective selection. There were cows for sale all over the place, but I wanted good cows. I turned to the books for advice. When establishing a herd of cows, wrote one authority, go to farm dispersal sales and buy all the oldest animals that you can find. If the farmer has held on to her for a long time, it means that she must be a first-class animal and therefore well worth buying. To confirm her quality, check on the size of the milk vein. That seemed to be fairly logical; the larger the milk vein, the better as it meant that lots of milk had passed down it, but I was a little vague on the exact whereabouts of the milk vein. Did 'large' mean the diameter of a drinking straw? Or a garden hose? Or a car tyre inner tube?

I turned to the other authority for confirmation. When establishing a new herd, always buy heifers. By doing that you will not be buying stock that someone else has discarded. You will not be buying in another farm's diseases and you will be able to mould the animals to suit your own habits and routines and not upset them by having to break those that they have built up over the course of their lifetime. This conflict gave me an inkling of one of the great truths of agriculture. Everybody who has any connection with the industry is an expert, and every expert has an entirely different solution to every single problem.

Realising this did not help me with my difficulty. My sublime ignorance would surely mean that I would be taken for a ride at every auction I attended and would end up with a bunch of highly expensive screws. I ran round the neighbours, shouting for help, and discovered another great truth and delight about agriculture. Unlike every other business with which I had been involved, the practitioners do not look on each other as competitors but as allies. The deadly rival against which farmers fight is not the farmer down the road who is undercutting them but the twin common enemies of the bureaucrats and, far more important, the weather.

Since each of my neighbours, as has already been established, was an expert, they were only too delighted to swamp me with totally conflicting advice and were particularly interested as they all had shares in the local sweep on how long I would last before I went bankrupt.

I finished up with the local auctioneer and poured out my little problem to him. His effectiveness at his job lay in his intimate knowledge of all the herds and virtually all the cows in the county and in being able to recite their pedigrees and milk yields unto the fourth generation. I stayed with him. He leaned towards the heifer-buying theory and he undertook to supply me with stock that he knew and on which I could rely. I considered this and found it good. I would buy through him using his experience and prestige. If he bought me rubbish, then I would simply not buy from him again. If his selections suited my requirement, which was to make me a fortune, then I would continue to buy from him in the future. Hopefully the incentive of future business would keep him on my side.

He supplied me with about thirty-five in-calf heifers, about three months off calving. They came from two herds, both with above-average records of milk production, and the gap between their arrival on the farm and their calving dates would give them time to settle in before I attempted to milk them.

The animals arrived and were decanted into the fields and meadows. I proceeded to make friends with them, the theory being that if I spent long hours with them, they would come to regard me as harmless and probably friendly and would work hard to do my bidding. I bought a selection of numbered plastic collars and draped them tastefully round their necks. I thought about christening them with traditional cow's names like Buttercup or Nellie but decided that I would manage to insert all the rage or love that I should ever require into plain-numbered 'four' or 'thirty-two' or whatever. A very clear precedence emerged. Number four was the herd boss, demanding the best lying place and first grab at any titbits that might be offered. From her, the precedence worked all the way down to Number six who was the general dogsbody and was chased around and beaten up by any cow that could not think of anything better to do.

Number 6 was the patient in the farm's first medical emergency. She had seemed to be a little simple-minded anyway, but now she appeared to be getting thinner and, instead of eating, stood and drooled at me when I inspected the herd. I cornered her and called in the vet. This was the first time that he had put foot on the farm and he strode up to the patient, shoved his arm down her throat and pulled out yards and yards of tongue. He indicated a raw patch at its base, slapped on some disinfectant and disappeared again in a whirlwind of plastic coats, gloves and scrubbing brushes, with his car radio blaring out commands for his next call. Both the cow and myself were rather overwhelmed.

Just when I was beginning to become rather uneasy at the imminent approach of some calves and the trials of the subsequent milkings, one of the neighbouring experts decided to change his parlour and needed some help for a few days while he pushed the cows through a temporary structure until the new parlour should be ready. This was a godsend. At last I was able to get my hands on a cow's udder and get some idea of what one was supposed to do in order to extract milk. I was not allowed to do much handling of his precious cows, but at least I could really study how it was supposed to be done. I only managed two milkings before my own lot started to calve. Then I was in business.

# Chapter Two

The milking parlour is the sharp end of dairy farming; it is the equivalent of the Western Front during World War 1. In theory, it is where the farmer in his spotless white coat cheerfully extracts milk with gleaming equipment in a well-lit antiseptic atmosphere from a herd of glossy contented cows. The whole should be bathed with the warm, rich sounds of Radio 2 at its most saccharine.

My start had a fair bit going for it. The parlour was clean and freshly painted, as was the second-hand equipment. I even bought a new radio. Granted, the walls were damp, the paint soon peeled off and the rust quickly showed through on the old ironwork but, in general, the parlour's image was not too divorced from what it should have been. The theory demands an operator who knows what he is doing and here the theory and the practice parted company.

The first of the heifers to produce was Number 23. She subsequently turned out to be a cow with real qualities of character, with an amiable, unflappable disposition and was bloody enormous to boot. However, such discoveries lay several months in the future. I had not quite finished the parlour when she calved. I was taking my evening stroll amid the herd looking for signs of calving, death, rotting tongues etc, when I espied 23 standing by herself in a corner of the hedge. She had had herself a calf, which looked to the uninitiated eye like an Aberdeen Angus.

It was a thrilling moment. What should I do with the thing? I raked through my memory of the college notebook and clicked into action. The first thing to do was to separate 23 from the afterbirth. It had come out all right, but she was busy trying to eat it. This is one of the little habits which cows have that need to be discouraged. Quite apart from the revolting shlurping noises she was making that had to be heard to be believed, my understanding was that the afterbirth could choke the cow. I grabbed it, in texture rather like half a hundredweight of frog spawn, and enjoyed a brief tug-of-war with 23 before I won and, to her outrage, tossed it over the hedge.

Next problem was to bring in the calf. It was lying down looking sweet and had obviously been born several hours earlier. 'Here, calfie' did not get any response, hardly surprising in a creature so young. I approached and, resolutely resisting the temptation to coo, patted it smartly on the flank. It blinked its eye enigmatically and lay as if paralysed. Panic. Perhaps it was paralysed. I had to break off the encounter to extract the mother from the hedge where she had got stuck while trying to retrieve her disgusting snack. I returned to the calf. I bent down and heaved the thing to its feet observing, as I did so, that sexing calves should not be much of a problem. The calf teetered for a second and then collapsed, its chin hitting the ground with a resounding thud. It fell over on to its flank and rolled its eyes at me so that nothing but the white showed.

Was it dying? Should I summon the vet again? The mother was ignoring us and was half way through the hedge again. I bent down and lifted her offspring in my arms, hearing my spine creak in the effort. I was intent on getting it quickly into the buildings so that I could give it urgent attention. The calf convulsed in my arms, let out a bellow and peed down my front. It was a bull calf. With a heifer there is a good chance that she will excrete away from you. No chance with a bull. The bellow brought Mum back from the hedge in a flash. She stood looking wildly round her for the calf everywhere but at me. She then let out a moo and trotted round me in a circle. The calf let out another bellow and 23 skidded to a halt and peered suspiciously at me. Are cows dangerous in defence of their young? Lack of knowledge appeared as a mighty chasm at my feet. Discretion seemed the better part of valour and I carefully placed the calf back on the ground. The mother gave another snort and moved towards it. The calf rolled its eye at her in a terrified fashion and leaping to its feet with a wild bellow took off at an astonishing speed, legs all over the place, with myself, the mother and the rest of the herd in hot pursuit, tails raised, heels kicking in the air.

Three hundred yards later, the calf tripped and fell flat on its face. Immediately it was surrounded by the cows. I elbowed my way through the crowd feeling rather like an ambulanceman at a motor accident and grabbed the calf firmly by the tail. Again it leapt to its feet and tried to run off. It was like holding on to a high-pressure hose. The tail gave an ominous crackle. I dropped it as if it was burning and away the calf went again. Once more it ended up on its nose. Panting a bit, I caught up with it again and lifted it up. The temptation to coo had completely evaporated. This time things went comparatively smoothly. I lurched my way down to the buildings with the herd milling and bellowing round me. Eventually I reached the bull-pen next to the cubicles and dumped it unceremoniously on the floor. First problem out of the way; now to milk the mother.

The herd, which was still in a distinct stew, had obligingly come out of the fields into the yard so that they could better observe the tortures to which I would subject the calf. It was the work of an instant to slam the gate behind them and prepare to run them through the parlour and milk Number 23. The first move was to drive the herd into the collecting yard beside the milking parlour. I had not yet plumbed the depth of the cows' psyches and did not realise that anything out of the ordinary will tend to make them behave rather badly.

The animals did not want to go into the collecting yard. There was this fellow who used to wander amongst them and occasionally scratch their noses having the cheek to demand that they did something. They clearly thought it was outrageous. It was the first test of the collective herd will against my own. I got behind them and yelled and whooped. I summoned the wife and the dog and we all yelled and whooped and jumped up and down. The cows behaved as cows behave. Those behind cried forward and those in front cried back. One of the ultimate weapons in cow control came to the rescue. A fist-sized rock bounced off the backside of the cow that was acting as the cork in the bottle in the gateway and sent her rocketing through. A rise in pressure sent the rest suspiciously in her tracks and I gratefully slammed the gate shut behind them.

First step towards milking 23 was now completed. Already grim experience was teaching me how to be a dairy farmer. If the animals did not want to go into the collecting yard, how much less would they want to enter the milking parlour? The yard was strange because it had a roof, but the parlour would be full of hissing, clanking machinery and strange excrescences of pipe-work, let alone the manic tones of Jimmy Young.

I flung open the door from the yard to the parlour and from the pit optimistically raised my voice in summons. There followed a frantic scrabble of hooves and total silence. I went to investigate. The animals were all piled three deep against the far wall of the yard regarding me as a group of novice nuns would regard a flasher who had invaded their convent. My heart sank. I attempted in a half-hearted fashion to winkle a cow or two out of the pile and lever them towards the door.

When pressed, a cow can form a circle with her mates that gives her almost total protection. Instead of arses to the middle and heads out to drive off the attacker, she sticks her head in and her behind out. I was presented with a solid row of backsides which pressed firmly together as I tried to squeeze between them. The only way to guide a cow is by heading her off. Tailing her off is impossible. Faced with this immovable phalanx of behinds with the attached heads five safe feet away all craning back to look at me, I recognised defeat. Cunning would have to succeed where firmness and leadership had failed.

So there the animals were left. The only way out was through the milking parlour. Twelve hours later they were still all there. The only difference was a considerable quantity of slurry now sloshing around. Eighteen hours was the cracking point. Hunger began to outweigh terror and, one by one, they went through the parlour in a scrabbling rush.

This was progress of a sort. The beasts were prepared with more or less persuasion to come into the milking parlour. The next point was to intercept and milk 23, at whom all this fuss and trouble was aimed. The parlour was in functioning order. Technically it was a 4/8 herringbone, meaning that four cows stood in the stalls on either side of the pit and there were four glass jars along with the milking equipment running through the centre. From there, the milk ran to the 250-gallon bulk tank in the dairy. After the first couple of times through, the animals were beginning to become accustomed to the sound of the pumps and machinery and were lingering to sample the concentrates that I was dropping down into the mangers from the cake loft above. Number 23 had been giving all her milk to her calf but now I

24

thought I would try to divert some of it my way.

She came in, walked calmly to her manger and started to eat her cake. I ran through the checklist that I had been mugging up. Wash tits. Strip foremilk in order to check for disease. Put on cluster and stand back to watch the money roll into the jar. I approached with the warm-water hose and sprayed it gently onto her swollen udder. She shifted uneasily from foot to foot — from large foot to large foot. With considerable caution, I stuck out my hand and grasped a tit and prepared to squeeze out some milk. Her hoof rose like lightning and flashed out, grazing the end of my nose. I backed off to the far side of the pit feeling a little weak at the knees. Clearly 23 did not like being checked for diseases. I would try her with the cluster.

At arm's length, I pushed a cluster between her legs and cautiously slipped the first shell on her teat. She languidly raised a hoof and brushed it off again. This was ridiculous. 23 did not appear to be aware that being milked was her destiny. This time, moving a bit quicker, I managed to clip two shells on to her before her hoof moved. She now became a bit more vigorous. The cluster was ripped from my hand, ripped off its rubber hoses, skidded through a dung pat and ended up in the corner of the parlour. 23 raised her head from her concentrates and gave me a supercilious look.

I thought for a minute that I might have to get tough. It was not much good keeping a herd of cows if I could not get any milk from the bloody things. I retrieved the cluster and carefully washed it down and then got out of the pit and collected several yards of that farmer's solution to all difficulties — baler twine. I proceeded to cocoon 23 in it, trussing every extremity to pipes, troughs and anything else that projected sufficiently to get a loop round. When she was secure, I returned to the cluster. I slipped it on all four tits. 23 heaved like a berserk eel but her bonds held. Approximately one pint of milk came out. I took the cluster off and spent ten minutes untying her. She cleared the front gate at the end of the parlour like a gazelle as soon as her last bond was free and disappeared into the middle distance. As I used the regulation fifteen gallons of boiling water to clean the equipment down, I pondered on this method of earning a living.

At 5 am, the time that all farmers start milking if they are new to the game, I went out into the fields to retrieve the herd and bring

25

them in to be milked. Horrors. Number 2 had also dropped a calf and now she, too, would have to be milked. The herd was now beginning to realise that it was part of the routine to come into the parlour and there was likely to be handful of caviare-like concentrates waiting for them inside, so there was a certain degree of muted enthusiasm to come in. I brought them in, leaving the calf out in the field for the time being. Numbers 2 and 23 were in the same batch. I spoke soothingly to 23 and wondered precisely what I was going to do. I washed her nervous titties and gave her an extra pull of concentrates. As she dug into them, I whipped on the cluster and rushed down the line to deal with 2. Before I had a chance to touch her, off clattered the cluster. I conceded temporary defeat.

I went up to the house and prised my wife away from a pile of nappies and brought her down to the parlour to help.

'I want you to stand up beside Number 2 and keep her mind occupied with something else while I put on the cluster.'

'How?', said my wife, looking rather dubiously at 2 who was looking nervously over her shoulder. Somebody else in the milking parlour was a potentially dangerous departure from the newly established routine.

'I've no idea. Use your initiative.'

She looked rather doubtful but moved up beside 2 and whispered sweet somethings in her ear while I tried to wash her udder. 2 put her ears back and loosed a probing foot in my direction. 'Try something else, dear.' The 'dear' through gritted teeth.

'Well how am I supposed to know what to do?' This was the nub of the problem. Neither of us knew what we were supposed to do.

'Try something physical. Grab her tail and give it a twist.'

'But she might kick me.'

'That's the whole point of the exercise. If she starts kicking you, she might stop kicking me and I might be able to get the cluster on.'

'I don't think I'm wholeheartedly in favour of the exercise.' Still she took hold of 2's tail and gave it a gentle twist. 2 grunted and loosed half a bucketful of slurry in the direction of her tormenter.

'God. How revolting.' My wife stepped hurriedly back and slipped on the wet concrete. She fell flat on her behind. While 2 was gloating over her triumph, I managed to slip the cluster on her. With contemptuous ease, she brushed it off again. My wife was not happy and stayed just long enough to tighten savagely one of the knots on 2's twine cocoon that she thought might have been a little loose. Then she flounced back to the house carrying with her a miasma of cow dung.

That day the total milk production of the farm was four pints and remarkably stupid it looked as it nestled snugly in the bottom of the bulk tank while the paddle that stirs and cools the milk spun vainly above it. I phoned up the dairy to find out what I should do with it.

'Just informing you that I'm now producing milk.'

'Fine, the tanker will be round at about 9.30 am to pick it up.' He sounded very efficient.

'One small point,' I said.

'Yes?'

'I'm not actually producing a great deal of milk as yet.'

'It doesn't matter. We don't expect everybody to fill up a tanker when they start. Ha ha.'

'Ha ha,' I agreed. 'Four pints?' There was a pause while it sank in.

'Did you say four pints?'

'Yes, well about four pints. It may be a bit more or a bit less.'

'Are you serious?'

'Of course, I'm serious. I have milked my cows and have four pints for sale. It says in my contract that you have to use your best endeavours to pick up my milk. So bloody well start endeavouring.'

'Yes.' A note of pleading entered his voice. 'But you can't expect the tanker to divert one-and-a-half miles and take up twenty expensive minutes of the driver's time just for four pints.'

'Well, what do you suggest that I do with it?'

'Drink it?'

We compromised. He could see that I was not going to be diverted from my determination to sell the very first milk that I had extracted. So I poured it carefully into a churn and lugged it to the top of the lane and he arranged for a lorry to come and pick it up. I was just in time to intercept a considerable amount of lip from the scornful lorry driver. In my euphoria at actually being in business, I was not deterred.

27

His revenge came a day or two later when I received a note from the dairy saying that I had failed the hygiene test. Apparently I was supposed to cool the milk after it came out of the cow and before it went into the churn. They might have told me. That was the first of many loving letters I received from the authorities on a wide variety of subjects, all threatening fearful retribution should any of my manifold offences be repeated.

During the rest of that week, I thought long and hard about the problems of milking. Even after only two or three attempts, I already felt a sensation of dread before I turned on the milking machine. The Monday morning feeling was one of the main things that I had left civilisation in order to avoid, and here I was experiencing it twice a day.

I started to protect myself. I found a stout piece of perspex and with the judicious addition of boiling water, I fashioned it into a curve and slipped the thing inside my trousers and sweater. I now had body armour from my neck to thighs with a vital tongue of perspex going even lower. Stout rubber gloves and a crash helmet completed the gear, and I approached the cows with some degree of confidence.

This was rapidly dispelled. The calves were falling like autumn leaves and there was a whole new crop of homicidal heifers to be milked. The pain of the kicks was lessened by the armour but it was the unexpectedness and the impossibility of anticipating them that frayed the nerves. I could not tie up every single cow every time. I would just have to persevere.

I cracked one morning when 23 came in. She, who should have known better, drummed away a tattoo on my armour. The blow that broke me caught the tongue of perspex and gave me a distinct twinge, although it was the thought of what might have been rather than what really happened that did it. With a shriek of fury, I buried my fist into her belly. I think it is true to say that this was the only occasion out of the many scores of times when I used violence on a cow, that it actually had the desired effect.

The cow is a very knobbly, bony beast, full of agonising fist-bruising projections and completely encased in a stout suit of untanned leather. The only vulnerable point is the udder, and that you dare not touch because from that flows your living. Punching 23 hurt like hell but as a relief to frustration, it was bliss. 'Right, you miserable cows (it is one of the tragedies of dairy farming that calling a cow a cow does not have the same impact as it has if you use the word on a traffic warden, or even a wife), you lot have given me a hard time and I will give you a hard time back'.

With my new-found determination, I thrust the cluster on to 23. No more cowering away from her so I shoved my shoulder into her flank. I felt her muscles tense as she prepared to kick and, it still embarrasses me to think about it, I sank my teeth into her side. My hands were busy with the cluster at the time. She relaxed and the cluster stayed on. With each succeeding cow, my surge of adrenalin carried me through, swopping punch for kick. During that first few weeks, my knuckles swelled and even now, five years later, I still have one knuckle that is twice the size of its opposite on the other hand.

It was a brutal uphill struggle. In my massive ignorance I assumed that everybody's milkings were like that. Looking back, my mind boggles. Even the experienced farmer will quail slightly at the thought of breaking one new heifer into the routine of milking. I had blithely brought an entire herd of them into premises that they had never seen before, to face a milker who had never done it before. If I had known anything about it then, I would never have contemplated trying it.

At the end of the first month, things suddenly seemed to fall into place. Milking, from being a bloody battle, became easier. Kicking me or the cluster became the exception rather than the rule. There came the never-to-be-forgotten day when the milk recorder came for the first time. You have to pay for his presence and he takes samples of milk from each cow to test for quality and throws back a computer print-out giving details of yields and stage of lactation. He was new to the job and did not know cows. With blithe confidence he came down into the pit with me so that he could do his sampling. Once again it was a break from the routine that the animals were becoming used to and they showed it.

He crackled and rustled his way up the pit in his waterproof garments. 'Bit draughty in here', as the wind from the fractionally missing hooves stirred his garments. Another hoof blurred out. 'Don't be so frisky, dear.' He leaned underneath the cow, between her legs and scratched her stomach. I closed my eyes. When I opened them, the cow was craning round and looking at him with amazement.

'I wouldn't get too close if I were you. You know that cows can kick.'

'Nonsense.' There was a good solid thud as the cow recovered and caught him full in the stomach. He reeled back and sat on his box of sample jars and moaned quietly. At least the cows did not have it in for me alone. I did not blame the man but he ruined the milking. His cool was blown completely and every time an animal moved, he reared like a startled horse and the nerves of all the cows jangled and twitched in sympathy. At the end of the milking, he scurried out of the parlour and a very late tackle from the last cow sent his sample jars spraying from his arms and all over the place.

The cows, in spite of in-parlour disagreements, never seemed to hold much of a grudge against me. After a month or two, they were beginning to show that they had very distinct personalities. They had a very definite order of entry into the parlour. Number 4 was the first cow into the parlour. It is like a litany — considerably more effective than counting sheep if you are an insomniac because you have to make the mental effort to put a face to each number: 4,8,2,46,9,12,14,18. It does not make particularly interesting reading.

Milking characteristics varied enormously. If you waved the cluster within a couple of feet of 8's tits, the rush of milk could sweep you out of the parlour. It was a question of how quickly you could get the machine to her to prevent her squirting the farm's life-blood all over the floor. Numbers 18 and 6 were the opposite. A cow has to be in the right frame of mind before she will let her milk down into the machine. Number 6, an elegant black animal with an elegant black udder, was always too hen-pecked by her herd mates ever to relax sufficiently to be milked properly. Number 18 had teats that stuck out horizontally from each corner of her udder. She was a cow full of good intentions but, poor animal, however hard she tried to help, it would take twenty minutes to empty her udder rather than the more normal five.

All the cows had their little idiosyncracies. Number 1 always refused to stand in the front stall of the parlour. This would have been quite acceptable had she not also been exceedingly greedy. This led to her standing in the doorway of the parlour wearing an expression of agonised frustration when a new batch of cows was called in. She wanted to get at the food but she was damned if she would go in first. Complete stalemate, as no other cow could get past her. I would hurl cloths and sponges at her until she backed away and let another through. She was never a great kicker in the true tradition of great kickers, like some of the rest of the herd, but she was a confirmed tail swisher. That would not have been a cause for concern had she not been rather dirty in her habits. She could not be bothered to lift her tail out of the way when she had a pee. I would come to her and wash down her udder and inevitably she would wrap a yard or two of wet and revolting tail round my neck.

Hygiene in the parlour was always a problem. Cows scatter dung about with a merry profligacy and irresponsibility that can only be equalled by a politician making his or her promises. The subject of dung is of absorbing interest to farmers, not because they have some sort of bestial anal fixation, but because so much time is spent in avoiding it, disposing of it and generally shifting it around. In the milking parlour one inevitably came across the stuff rather more brutally and immediately than one would have liked. I stood with my head 18 inches or so below bum level and,

31

with a full complement, all eight bums would be pointing inwards.

Excretion of both varieties is the cow's way of expressing emotion. Normally things were not too bad, there were just some gentle splashings on to the floor of the parlour with consequent dispersal of spray. A determined pisser could arc out into the centre of the pit and some neat footwork would be required to avoid a refreshing warm shower. Worst of all was the rare but lethal combination of a raised tail and a cough. When that happened, whatever the cougher had on offer came out with the force of a shell from a gun and woe betide anything in the way. The couple of occasions that I received a shot full in the face remain indelibly imprinted on my mind.

It was the constant spatter of dung during milkings that provided the real problem, particularly when the animals were young and rather emotional. A judicious kick on the cluster would mean that it would come off, with each shell sucking frustratedly at air, deprived of its usual meal of milk. If the cluster should happen to end up on a cow-pat, it would seize on it with a slurp of delight and large chunks of the pat would disappear into the network and I suppose would ultimately end up on a doorstep in SW1. Of course, I would try to avoid this distressing eventuality, but when only a little bit of dung went in and there was already an awful lot of milk in the jar, it could be very difficult to pull out the plug and pour it all down the drain.

Apart from the ever-present threat of the loose back-end, there were various other ways that cows could express themselves. The most usual was the hearty boot in the belly. Subtle practitioners could achieve the indirect boot in the belly, if I was out of range. This entailed kicking the cow next door so that she would lash out and catch the aforesaid belly. Other methods included the manger rip-off. If a cow was in a bad mood or felt that she deserved more concentrates, she would rear up and put her front feet into the mangers which were bolted to the parlour wall. With a bit of enthusiastic jumping up and down, the whole lot would come cascading off the parlour wall and take days to replace. The herd was milked twice a day and so there was never sufficient time to allow the repair cement to dry off and so down they would tumble again at the next milking.

Most of the cows soon learned to come into the parlour from the collecting yard without much persuasion and would wait patiently behind the chain which prevented access to the mangers until I let them in. They would peer thoughtfully down into the pit at me and if I came within range would lean down and grab an enormous tongueful of my hair. Those with a more logical turn of mind would at least once in their career decide that the concentrates could be more easily got at if they came to join me in the pit, and I would have to belabour their snouts until they retreated.

Once milked, there was a straight exit for them back to the fields or the silage pits. They did not always avail themselves of the opportunity. On wet days, the back of the parlour would clog with cows reluctant to go out and face the tempests. Number 8 would always stay behind and stare at me throughout the milking with a rather infatuated expression on her face. 8, bless her heart, always thought that I was marvellous. During the boring bits of milking when peace reigned with all the clusters sucking milk, I would perch myself on the end of the pit and she would amble over and stretch her neck out above my head and demand to be scratched. It was all very flattering and disarming.

The exit part of the parlour was also the main stamping ground of the bantams. These birds were obtained during a moment of extreme thoughtlessness because they looked pretty and might give eggs. They laid eggs all right and I would come on little piles of them stacked up in odd corners of the barn. One never knew how long they had been lying there, and after one or two disgusting experiences, I took to feeding those with an uncertain pedigree to the dog. Every few months, a clutch of these eggs would suddenly hatch and anything up to fifteen chicks would cheep their way round the yard, falling into puddles or being trodden on by cows. They always seemed to reproduce themselves at precisely the right rate so that·their adult numbers never dropped below eight or rose above ten, death carrying away the excess.

The birds reached their daily nadir at about 6 am during the morning milking. They would wake up when the machine was switched on and would tip-toe into the parlour to pick up any concentrates that might have fallen from the cows' troughs. I would usually be sitting, comatose with sleep at one end, while the cows would be rocking dozily in the rhythm of the milking

33

machine, until the cock's shrill clarion in the confines of the four walls of the parlour would almost blow our heads off.

They thought it their duty to test the content of the mangers before each batch of cows came in. As the milked batch left, there would be an undignified scramble of bantams and they would all hop in, cocks to the fore. Then the next group of cows would charge in and there would be a brief flurry of conflict before each manger would erupt its squawking bantam which would retire to the back of the parlour to await further opportunities. Regularly, myself or one of the cows would lose our tempers with these birds and would either hurl a handy object in their direction, or drive them out with much stamping of hooves and snorting, depending on species.

Number 23 was the only animal that learned to channel the drivelling uselessness of the bantams into a worthwhile occupation. One of the smarter birds had discovered that if it hopped on to the back of the cow in front of the line, it could then progress upwards over the pipework to a beam beside the cake loft. From there it could push its head through a hole and help itself. 23 made the Newtonian discovery that if she banged the chute that dropped the cake from the loft into the manger, the bantam took fright and would squawk and flap its way to safety, dislodging a worthwhile quantity of cake. From then on, as she left the parlour, she would always look up to see if the bird's back-end was poking out. If it was, she would give the chute a thump; if not she would walk on out.

Number 11 provided the emphatic full stop to that first period of being a dairy farmer. She always had been a pig of a cow with a kick like a steam hammer. I was quietly day-dreaming about the future pleasures of retirement one evening when I brushed her with my shoulder. She lashed out and caught me on the forearm. It caused one of those pure, exquisite agonies that come but rarely. It was even better the following morning when she managed to jar it again when I was putting on the cluster, and I had to pause to barf up my morning coffee on the parlour floor. I took myself along to the local hospital and returned home a couple of hours later with my arm in plaster. Sweet 11 had succeeded in breaking my arm.

In a sense I was grateful to her because it led to my education in how to milk cows. I was faced with an urgent need to find a relief milker since my punching capabilities were severely impeded by the plaster. I found one and for a couple of months I sat and watched how an expert milked and handled a herd of cows. It was a revelation.

During that period I enjoyed a desultory correspondence with the Department of Health. *Describe the nature of the injury for which you are claiming benefit.* Broken arm. *How caused?* Kicked by a cow. *Day of injury* Wednesday. *Day when stopped work* Thursday. *Please confirm day of injury and day stopped work.* I confirm. *You seem to be in error. You claim that you worked the day after the injury was received.* I do so claim. *Why?* Because if I hadn't milked the bloody cows, nobody else would. *How did you work on Thursday with the injury you claim to have received?* Carefully. *According to our information, it would not have been possible to work on the Thursday with the injury so described.* It would have been possible if you were as brave, steadfast and so full of the grit and determination so characteristic of our island breed as myself. *Herewith a cheque for £12.67 for benefit for the week commencing Thursday . . .*

# Chapter Three

It dawned on me rather later than it should have done that the cows were supposed to produce a calf each year, so that they could continue to yield the milk from which we made our living. Soundings established that the recognised method of producing a calf was through artificial insemination. To have a cow inseminated was easy enough. It only required a telephone call to the local AI Centre and the expert would come winging. I often used to wonder what they put on their passports in the spot marked 'occupation'. And as for those who actually did the semen-collecting from the bulls . . . My difficulties started when it came to deciding when the cow was ready for action. The herdsman has to catch the cow during a 12-18 hour period during her monthly cycle. Only if she is then inseminated will she become pregnant.

With some of the cows, it was not too difficult. Bulling was the giveaway, when one cow would ride on the back of another. But sometimes the animals would bull when neither of them was on heat, or bull when I was not around to see them, or even not bull at all. For some reason they always seemed to bull at the weekends, which was not very popular with the inseminators. The conversation would run something like this.

'Good morning, sir.'

'Good morning.'

'You think you've got another one on heat for me?'

'Yes.'

'Is it the same one that you thought was on heat last week?'

'No,' I would say indignantly.

'Oh no. I see on the certificate that it is 22 this time. You've got her down for a repeat service again. You realise that every time we have to repeat a service it reflects on us?'

'Yes, but I can't help it if the cow keeps bulling.'

'Of course not, sir. But you can see that if she really is on heat and she doesn't hold her service, it might look as if I am not much good as an inseminator.'

'Oh, I'm sure you're one of the best cow-stuffers in the business.'

'Very delicately put, sir.' He would have his polythene glove on by now and would have pulled the semen straw from its bed of liquid nitrogen and be holding the long insemination tool in his mouth.

'You got me down to do 20 bloody 2, six weeks ago.' This would be muffled by his mouthful. 'And a cow only comes on heat every four weeks.'

'I'm sorry. But I can't help it if I saw her bulling yesterday.'

'You didn't see her bulling, sir.'

'Didn't I?'

'No sir. You stick your arm up here and you'll see that she's six weeks pregnant.'

'Oh, well done.'

'You realise I'm going to have to charge you for a completely wasted trip sir. Why don't you buy a bull?'

The one advantage of AI was that you could use first-class Friesian bulls to father the offspring of your cows. If I took his advice and bought a bull, I would lose this. Any Friesian bull that I could afford to buy would have no chance of producing daughters of the quality of those sired by the highly expensive aristocrats that filled the insemination centres. Because my problems stemmed from my inability to identify heats, it would mean that I would have to give the bull unrestricted access to the cows. All I had read and heard about Friesian bulls led me to believe that it would be safer to employ an animal straight out of the Madrid bull-ring.

Already I was coming to the conclusion that the cows did rather better the less I interfered with them and tried to manage them. It was a hassle both for myself and the cows when I trapped them in the pen for the AI man's pleasure. And hassle was something that I wanted to avoid, in order to conserve what small management ability I had and expend it in areas that would show a profit. If I bought a beef bull rather than a dairy bull, that would cut out great areas of hassle. The calves that would result would be sold as beef calves and I would not have to become involved in the esoteric arts of rearing my own heifer replacements. If I wanted more cows (I was up to fifty by this time), they would be bought in at the most profitable time, as adults. The acres that we had would all support money-yielding cows rather than money-consuming young stock.

That's how Joe came to the farm. He was a Hereford, not because that was the best breed for the purpose but because one of our neighbours reared Herefords. He was also very cheap because he was small for his age. This, we were persuaded, was an advantage as our cows were not particularly big and the weight of a mature bull might well drive them into the ground — particularly on a farm as boggy as ours.

I went round to collect him. A piece of string was placed round his neck; I led him down the farm lane to the house and everybody came out to coo over him and pet him. He was then released into the fields and immediately proved himself by hauling himself and his stubby little legs on to the back of the nearest cow and pleasuring her.

Joe really was and, last time I heard, still is a bull of enormous gentility. I was kicked and bullied so many times by the cows that I always kept a slightly wary eye on them. With Joe, I never had a moment's unease. There were perhaps a couple of occasions when we misunderstood each other but nothing serious. After he had been on the farm for a week or two, he did try to mount me as I was bending over to put some hay in a trough but this was just a moment of thoughtlessness rather than a sexual aberration. It was not a consummation devoutly to be wished.

The first time that he felt frisky was slightly alarming. Cattle sometimes become carried away with the joys of living and take off in a series of breathtaking high kicks like young lambs; ponderous old matrons with pendulous udders or heifers, they all do it. The first time Joe performed like this was when he skipped and snorted round me for five minutes until he had the courtesy to desist when he saw the stark terror in my eyes.

In later years, he grew larger and considerably more portly. In maturity he looked a really first-class bull. One of the most vivid pictures that I have retained from dairy farming was from another occasion when the joy of being alive overtook his more conventional behaviour. He took off against the skyline with ponderous dignity, rather like a skittish road roller. With every high kick, a great roll of fat would slither down his body to end up round his neck, the same sort of moving ripple that you get when you shake a rug. This would cause him to lose his balance and he would teeter dangerously until he regained equilibrium. Then he would kick the other way and it would slosh back down to his tail.

Joe turned out to be exceptionally good at his job of getting the cows in calf. The average bull tends to fall in love with the cow of his choice and will pleasure her until her heat period is over before moving on to the next cow. Joe got his kicks out of variety. He would ride any and every cow that was available with commendable efficiency. He had trouble with one cow — 43 — who had a touch of lanky Holstein blood in her. His stubby little legs and her great gangling stilts created an impossible situation and I was nearly forced to turn her into meat pies until he discovered the trick of serving her on a downhill slope.

One of his crosses in life was that the herd always tended to calve at the same time — within six weeks or so of the beginning of September. Although his efficiency brought the herd-calving date forward a month or so, for a good half of the year Joe was considerably underemployed. This was in spite of the fact that I continually fed in autumn-calving heifers, whenever replacements or additions were needed. He would have a wonderful couple of months when he would shed his paunch and roam the farm with a manic glint of lust in his eye and just as the bags were beginning to form under his eyes, the source of willing flesh would dry up and he would find that his wives were all in calf and would spurn him. During his most lusty years, he solved this particular problem by diving through hedgerows and raping bunches of decidedly under-age heifers belonging to a neighbour. This was highly unpopular both with the neighbour and the insurance company. The claim form included a neat box entitled 'sketch of incident'.

Latterly he switched his emphasis. If he could not get his end away, he would just suck udders as the next best thing. Not any old udders but just the udders on one particular cow. The textbooks were rather thin on advice on how to stop bulls behaving in this fashion. It is supposed to be a problem that occurs with freshly weaned bullocks and the odd rogue cow, not with virile great bulls. He completely milked out the cow of his choice. When she calved and was in the full flush of milk, I might be lucky and manage to extract a gallon or two out of her. Then Joe would get this eye in properly and that would be the end of her lactation as far as I was concerned. He must have cost £400-500 a year in lost milk.

We tried various methods of preventing this. One took the form of an aluminium disc covered in spikes pointing outwards that was placed in his nose, the theory being that the cow would not like her udder being jabbed as he slaked his lust and would boot him in the chops. Putting this spikey moustache into his nose was no problem. I just walked up to him in the field and slipped it on while he blinked his plummy eyes at me. It did not work. The first thing he did was just jab at her. She twitched a bit, grew junkie-like punctures on her udder but she was prepared to tolerate it. I took off his moustache and sharpened the spikes. Joe got kicked. I held my breath in the anticipation of success. He then stroked his nose carefully down her flank, thus tipping the disc up against his nose and out of harm's way and settled down to suck.

The next attempt involved carefully painting bitter aloes over the cow's teats. Joe wrinkled his lip in disgust and kept on sucking. Next was covering the end of her teats with a skin of plastic dressing. Joe sucked until he turned blue and broke through. The final effort was clothing the cow in a brassiere made out of an old sack. She took a very dim view of this, as did Joe, and together they would work it off. I persevered with various fiendishly subtle knots and loops until it was secure and so Joe shrugged his massive shoulders and moved on to another member of the herd. At that point I gave up and let him suck.

One could make out a case stating that it would have been more sensible to dispose of a bull that cost so much to run, but I considered that his one aberration was more than made up for by his amazing efficiency. In any event I became too fond of him to think of sending him to the knacker. He would always stick with the herd at milking times right down as far as the collecting yard. If there was nobody on heat, he would position himself ouside the yard and wait. If there was a fruity damsel about, he would enter the yard and go about his business.

At the end of milking, I would release him from the yard again.

It was not entirely straightforward. The procedure was that I had to open the wicket gate and stay beside it. Joe would amble half way through and stop. Then I would have to give him a scratch. Not an ordinary doggy type tickle but a full-blooded enthusiastic rake preferably with a screw driver or similar pointed instrument. This would elicit clouds of reddish hair and grunts of delight. The operation was completed by a resounding slap on his rump that would send shock waves rippling up and down his layers of milk-fed fat and then he would stroll out. Open the gate and fail to scratch him and the yard would resound with his high-pitched bellows of protest.

There was one day when Joe got it all a bit muddled. He stayed out of the yard and clued up too late that he should have gone in because there was a cow just begging for her conjugal rights. He paddled round to the parlour exit and stuffed his enormous head through the door. This was not that unusual as he liked to keep an eye on things. It usually resulted in a missile being hurled at him as he acted as a cork, preventing cow egress.

On this occasion something turgid was obviously meandering around inside his pudding-like brain as he emitted one of his choir-boy bellows. I was concentrating on a cow at the other end of the pit and unable to intervene. He ploughed his way through the crush of cows, 8 waiting to lick me and 23 waiting for the bantam to perform and came to the edge of the pit. He looked in, came to his decision and down the steps into the pit he came. In some situations you do not argue with a bull, however friendly he is and this was obviously one of those situations. The entrance to the pit was narrow. It was meant to allow me in and keep the cows out. The tubular metal was not, however, designed to stand up to a ton of determined bull assisted by gravity as he came down the steps. They gave way feebly. Once on the floor, he kept on coming while I wondered wildly if this was my cue to abandon ship. The odd cow was now aware that something rather unusual was going on and a couple lifted their tails to shower him with dung.

The centre of the pit was filled with a couple of hundred pounds worth of glass jars, but he delicately stepped along one side while I prudently kept to the other. He started up the steps at the other end and I could see that his resolution was beginning to fade as he was faced with the ironwork there. I was faced with the horrid prospect of having to share the pit with a bull during the rest of the milking and God knew how I was to have got him out at the end. So I gave out my best cow-driving banshee yell and sank my gumboot into his balls. It did the trick. His bellow went up the scales even higher and he catapulted himself up the steps. He then turned and, to show that everything was still in working order and why he had come in the first place, he fell upon the cow being milked at the end of the line and ravished her.

Joe was quite often responsible for leading the herd on exploratory expeditions round the surrounding countryside. For a farmer, quite the most heart-stopping moment comes when he goes down to milk the cows and finds that there does not appear to be a cow on the farm. They have vanished. To me it was always a source of perpetual wonder how cows could disappear so completely in such a short space of time.

The first time that they did their disappearing act was a month or two after I had started milking them. At crack of dawn I went out to the fields to bring them in and not a cow could I find. In mounting panic I toured the nooks and crannies of the farm disturbing only the odd fox or rabbit out for their morning exercise. I returned to pick up a car and toured the neighbourhood in ever-increasing circles for two hours, searching desperately for traces of dung on the road and enquiring the whereabouts of a herd of cows from bewildered poachers and anyone else who may have been about at that hour. I found no trace of them and came back to the farm to telephone the police to report their disappearance. They were all there, of course. They had been waiting patiently in the collecting yard for me to come to milk them. That was the one place where they should have been and it had not occurred to me to look there.

Usually herd escapes were not so straightforward. Of course, cows should never get out, yet I cannot believe that a stock farm exists that always manages to keep its animals on the premises. A moment's inattention will mean that a gate is not fastened properly, and I doubt that a hedge or fence has been built that can stand up to a really determined bovine assault. A mass breakout by our bunch was fortunately rare; not more than two or three times a year. Often the escapees would come running up the lane to the house and gambol all over the lawn and eat everything that grew in the garden. It is a curious fact of life that a cow will never move at anything faster than an amble under her own volition until she breaks out and then she will behave with all the restraint of a football hooligan and sprint away.

The cow's sprint is closely geared to that of her herdsman. It is another indisputable law that the cow will always travel very slightly faster than her herdsman during an escape. If you walk, she will walk. If you trot, she will trot. If you sprint, so will she at half a mile per hour faster than yourself. It is therefore useless to try to head a herd off as it will always end up at any given point seconds before you.

When our herd broke out and turned up the lane as opposed to coming up to the house, then the priority was to get back to the car as soon as possible. Usually the herd's excitement was such during the escape that the noise gave it away while it was still in progress. There were two exits from the farm. It was useless to chase the cows up the lane because of the law already described. One had to drive like the clappers out of the other exit accompanied by as many helpers as could be rustled up in the ten seconds available. At the end of the lane came decision time. The road through the farm bisected a square: round which of the perimeters of the square should we go? The idea was to meet the herd head on and force it into reverse. Gamble wrongly that the cows had turned left instead of right at the top of the lane and you could chase their tails as far as Cornwall one way and London the other. So we would speed through the lanes, gnawing our fingernails in suspense, in the hope that we would confront them before they reached another crossroads.

Surprisingly, only once in five years did we gamble wrongly and that proved to be a grim two-hour struggle of cutting through hedgerows before we overtook them. Generally we would meet the herd galloping down the narrow lane between the hedges towards us. The art was to stop a hundred yards in front of it and de-car. I would stand four-square in the middle of the road feeling a bit like Canute as the thundering multitude bore down on me. The helpers would fan out down the side of the lane and the cows would roll past them.

Joe, who would have almost certainly initiated the escape in the first place in the hope of a bit of spare tit, would be well back in the herd by this time, his stubby legs being unable to keep up with the rest. The herd would come skidding to a stop at the sight of myself leaping up and down in the middle of the road screaming insults and he would push his way through to inspect this obstruction and decide whether it represented a real threat to his continued progress. Having decided that I meant business, he would turn on his heel with a snort of disgust, whereupon the rest of the herd would follow suit and clatter off back down the road, soon leaving Joe toiling away in the rear once again. With a bit of luck the helpers would have got in position and Joe and his harem would be steered back down the lane.

Solitary escapes tended to be more wearing than those en masse. Joe was the most frequent and also the easiest to recapture as he was more trusting by nature than the cows, as well as being slower. When he went on one of his solitary searches for a touch of rape and pillage, I could usually capture him unaided. I would walk up to him and talk about the weather and corn prices as if he was the last thing that I was thinking about. When he was thoroughly engrossed, I would lean forward and grab him by his ring and haul him ignominiously back to the farm. Once his ring came away in my hand, which so bewildered both of us as being outside the rules of the game that he stood and let me fix it back on before I hauled him off. He sulked for a week after he had been rounded up by a neighbour on a horse. That was just too unsporting.

One animal nipped over the hedge to explore a neighbour's field on the grass-is-always-greener principle. She was a brute of an animal and was dry waiting for her second calf. She had nearly taken my head off during her first lactation by booting me at the base of the neck. When one is faced with a single cow in a field of fifteen acres with one ten-foot wide gate as the only exit, it is difficult to know how to go about extricating her. The tendency is to leave her to find her own way back, but not when you have been called out by the neighbour to remove her. It is rather like one of those tests of initiative given to aspiring army officers. The one way that holds out a real chance of early success is by introducing herd mates into the field so that you can bring her out as part of a bunch. This time the other cows were half a mile away and I was damned if I would risk further escapes by bringing them over. The other method that usually brings results is bribery through rattling buckets of concentrates or waving wisps of luscious hay in her general direction. I tried it, but the cow trusted me about as much as I trusted her and she kept well clear.

The only remaining method was coercion. Here I recruited five helpers. The condition of aid was that I had to surrender command of the operation.

'Right, then. I want you, you and you to spread out that side and you this side and we'll drive her towards the gate.' The cow had been keeping a wary eye on things and as we moved in she shifted up a gear or two from an amble to a brisk man-killing canter.

'Watch it . . . You dozy idiot, don't just stand there and let her walk round you. *Run. Yes faster. Go on. Oh for Christ's sake.*' She easily evaded the first attempt. Instead of coming at her in a chain, we adopted football tactics and attacked her at an acute angle with a series of sweepers covering any attempt at a breakthrough.

'Gently does it now. Watch her head . . . nice . . . nice. Watch it. Oh no.' The animal speeded to a sprint, her udder swinging like a highlander's sporran. 'Watch your right. Don't run straight at her, you bloody moron. Head her off. *Head her off. You useless pillock. Run. Yes you too. Not at me.* Oh hell. Get . . . out . . . of . . . it . . . you . . . bloody . . . cow.' The line was now in that state of exhaustion that only those who have chased cows can have experienced. That slightly-faster-than-you-speed goads you into a pace that you did not dream you were capable of, especially in gumboots. The cow circled easily behind us.

'OK we'll have another go. Get hold of a stick. No, you idiot, about six feet long so that it will extend your reach. Right then, move forward. Gently does it . . . watch it, Philip, she might go . . . *Watch it.* Oh well turned . . . *Get away you bitch . . . get away . . . you . . . miserable . . . bloody . . . cow . . . Take that.*' The boss, with an astounding spurt managed to break his stick, that common gesture of despair, over her disappearing arse before dropping to his knees and sobbing for breath. The cow, moving beautifully, plunged through the hedge into another field.

This was the crisis stage of the escape. The attack team was on its knees and was showing a lamentable lack of interest in continuing the chase, so I took control of the operation. With a persuasive eloquence that would have had Nye Bevan gasping with admiration, I forced them through the hedge after her. Here, temporarily, the cow had blundered. The field also contained a group of phlegmatic beef cattle. We managed to bunch the lot of them out of the field into the road and prepared to extricate the cow for the run home. When she saw what we were about she plunged through a hedge on the opposite side of the road and was away. We cornered her half an hour later in another farmer's yard and brought a trailer to carry her back home.

The rest of the team were unanimous. If that particular cow ever got out again, I could get stuffed as far as expecting any help from them was concerned. There was one exception. If I ever became tired of her and decided to send her for slaughter, then they would help me catch her and load her with pleasure.

That particular animal was young and whereas most of them, after a couple of lactations, settled down and became about as tame as lap-dogs, some of the young ones could create a mayhem that was quite awe-inspiring. Only one ever defeated me completely. She was one of a batch of five heifers that I bought a couple of months before they calved. She was small, dark and undistinguished and the whole batch appeared to be fairly house broken. She calved high up in a field near the lane and I trudged up to perform the usual chore of shepherding the newborn and the proud mother down towards the buildings.

It soon stopped being a usual chore. The mother was waiting quite placidly beside her offspring and I moved round to pressure them in the desired direction. When I came within ten yards of her, she raised her head, looked carefully at me for a few seconds and then dived straight through the hedge into an adjoining field containing the milking cows, rather than those in the maternity unit. Somewhat mystified by this performance as normally a cow needs a crowbar to prise her away from her calf, I decided to leave her alone and bring her in at milking time with the rest of the cows and concentrate instead on her calf. I gave it a gentle prod and the calf, a dear little heifer, staggered to its feet and charged me, giving my kneecap a painful rap. She tripped over a daisy and ended up as a pile of feebly waving legs.

I approached her rather more gingerly and gave her another prod. Gallantly she raised herself and came at me again. Normally one has to drive or carry a calf towards the buildings and it takes some effort. This animal was a dream, she came plodding grimly at me with a glint of manic aggression in her eyes. I led her straight into the buildings and clanged the gate shut as she was turning to charge me again.

The source of this astonishing display of ferocity became

apparent at the evening milking. The herd was ushered towards the parlour as usual with the calf's mother at the back but appearing to be perfectly happy. I moved towards her to ensure that there were no post-calving complications and blenched. As she caught my eye, she started to kick. All cows will kick given a chance and a bit of provocation. This beast was not being given the remotest provocation and yet her back hoof was thudding into the wall of the collecting yard with such force that the whole yard roof was shuddering.

She then came into the parlour itself like a lamb and started to wolf her concentrates — false alarm I thought and busied myself with the cow at the front of the line. A flicker of movement caught my eye. I looked and saw nothing. Another flicker. She was kicking at nothing, her foot scything out at head height, moving so fast that it was a blur. I moved closer and brushed a jar as I approached, with an awful fascination I watched her lift both hooves off the ground and lash out backwards like a donkey.

There are ways of dealing with a cow like this. The traditional kick bar, which is clipped over her haunches and is supposed to immobilise her kicking foot, is the easiest. In her case, as soon as she knew that her offensive capability was being threatened by the bar she would go berserk. Twice she toppled over the rail into the pit. Once and only once did she contact me with her foot and she hurled me back into a jar, smashing it and costing me fifty pounds for a replacement. With most cows, patience and tender loving care will eventually bring them round. A certain limited success was brought by conditioning. I held up a chunk of steel beside her foot and clicked my tongue. With each click her foot would lash out and hit the steel. After a dozen or so clicks, her kicking foot would be oozing blood and she was lame. Then I could put the kick bar on her and survive. I hoped that this would convince her that kicking produced an unsatisfactory return on investment but it was not to be.

I persisted with this animal for three months. She would be reasonably safe with the bar on for a day or two, and then with no warning at all she would lash out. The relief milker refused to go near her. Once a bantam got too close and intercepted a hoof which hurled it against the wall and broke its wing. That day it was a bantam, the next it might be me. I sold her at that point to a neighbour for her slaughter value — the only total failure that I

47

ever had. My neighbour passed her on to another dairy farmer to find out whether she was dangerous or merely hated me. She broke up his milking parlour and so he bounced her back. I was delighted; it would have been infuriating if somebody else had succeeded where I had failed. She stayed on the original neighbour's farm for another seven months until she calved again. Back she went to the dairy farmer for another try. This time she shattered his watch, gave him severe bruises and completely disrupted a milking. The following day her soul was in hell and her body on its way to make meat pies. It served her bloody right.

# Chapter Four

As a pig farmer, I had learned basic stockmanship. Transferring
knowledge gained from pigs to cows was not too difficult, as many
of the same basic principles applied. I was profoundly and
completely ignorant about everything to do with cultivation and
the tilling of the soil. Some farmers became all gooey eyed and
weak at the knees at the concept of The Land, ownership of a
small part of England and things. The Land, to me, was an
unpleasant but necessary adjunct to my prime business, which
was coaxing milk from a recalcitrant bunch of dairy cattle. It
always seemed to let me down when I needed it most by being
either too wet or too dry. It was dirty, smelly and cost a fortune to
run.

One of the most striking aspects of agriculture is that it
supports cohorts of highly qualified people all clamouring to tell
you how to do your job. Many experienced farmers consider them
to be a bit of a pain, but to me they were manna from heaven. To
them I was likewise for they could never have had a more attentive
and receptive pupil. The brief scuttle around the fields of my
neighbour in Wales had taught me little about how one was
supposed to treat the land so that it provided a constant source of
food for the cows, and so the first adviser was asked to report on
the peculiarities of the farm and how it ought to be handled even
before I had moved in.

His advice, I am sure, was sound but it was way above my head.
His report talked of pH levels and units of potash and S 23. When
he finally plumbed the depths of my ignorance, he became
splendidly dogmatic and laid down laws that should not be
broken. His dogmatism was larded with a profound contempt for
my presumption in thinking that I could make a go of keeping

cows. This became apparent when a file left on the farm a year or two later revealed comments that made my ears burn.

It was rather disconcerting to discover that all advisers were equally dogmatic and that most of them gave out total contradictory information. The usual procedure was for advisers to come to the house and settle down in the office with a cup of coffee. They would gently probe out my areas of knowledge and then with verve and enthusiasm launch into the gospel according to their employer.

The advisers were split into two types. Firstly were those employed by the Ministry of Agriculture and the Milk Marketing Board who were usually impartial but highly academic and sometimes hopelessly impractical. One of these types airily told me to bulldose all the buildings and start again, which would have been all right had I several scores of thousands of pounds lying fallow somewhere. The others were employed by firms with ideas and goods to sell me and they writ their messages large on the clean pure tablet of my brain. I was only saved from pressing my cheques upon them because I already knew that the next man would tell me to do something totally different. I was in danger of becoming like the centipede who became so engrossed in trying to work out which leg moved after which that he fell over and could not get up.

I was saved by an institutional adviser who decided to take me on a walk round the farm and discuss drainage and grass varieties.

'It is vitally important that the right sorts of grass are encouraged to grow. The ryegrasses, for example. Now this . . .' He stooped and plucked a piece of grass and rubbed it between his hands and peered at it closely. 'This is *Agrostis repens*.'

'Is that good?'

'No, that's bad.'

'I'm sorry.' He looked as if he expected me to apologise.

'Yes. Poor drainage and stocking encourages these weed grasses at the expense of the ryegrasses.'

'Really.'

'Yes, now this,' he bent and harvested another insignificant blade of grass; 'this, oh dearie me, is *Agrostis stolonifera.*'

'Not *Agrostis stolonifera*?' He looked at me rather suspiciously in case I was taking the piss out of him but I was not. I was just carried away with the atmosphere of the occasion.

'Yes. Definitely *stolonifera* and that . . . ' as we squelched our way through a particularly boggy patch, 'that is *canina*.'

'Do you know, it looks just like the other one.'

'Yes, I suppose it does a bit.'

'But how do you tell the difference?'

'The awns.' That shut me up. What the hell was an awn?

'What the hell is an awn?'

'I'd show you but I haven't got my glasses with me.' I did spot the deliberate mistake.

'But if you haven't got your specs, how can you tell one awn from another?' If I thought that I had caught him out, I was wrong. He was magnificent. With a disdainful sweep of the hand he dismissed such pettifogging doubts. 'It's all grass, just grass. If you can produce enough of it the cows will survive. If you can't, sell a few cows. No need to get your knickers in a twist.'

In the first year the grass, rather disconcertingly, failed to grow as it should. I completely omitted to put any fertiliser on because I was then working on the assumption that since grass seemed to grow virtually anywhere that was not constantly sprayed with weedkiller, it needed precious little help from me. I realised the error of my ways a little too late for much of that growing season. The other problem was the sheep. They had been on the farm far too long and once I had persuaded their owners to remove them, I had the devil's own job to keep every other sheep in the parish out.

We were in the centre of a sheep-growing area and there was a constant trickle of refugees popping through the most unlikely holes in the hedges to eat our precious grass. The faster that they were herded back and their points of entry blocked, the faster they found fresh holes through which to come in. I have now kept both cows and pigs and some of them can have their moments. Nothing on earth would ever persuade me to keep sheep. A more perniciously stupid animal was never invented by the Creator. Each autumn in return for huge sums of money we would be forced by local custom and convention to house some on the farm between November and Christmas in order to tidy up any remaining grass that the cows had left at the end of the season. It was a continual chore going round and extracting them from fences and ditches. All domestic stock has a deep-rooted ambition to die in order to spite their owners, but the sheep takes this suicidal urge to absurd extremes. If it finds itself on its back, it will

51

very probably die unless some passer-by happens to right it. The only other animal I know of that has the same talent is the tortoise and even then it very rarely actually goes as far as dying.

Herding these appalling creatures back through the holes from which they popped required a basic knowledge of the sheep's psychology. It is the same with all animals. You have to know the correct moment to yell and scream and jump up and down and the right moment to shut up.

The only time as a farmer that I completely lost my cool with a person as opposed to an animal was while herding sheep. For the third time in one day a dozen or so of the brutes broke into the one field that we had that was full of lush waving grass. I had a contractor on the farm helping to do some building work and we did the usual demented chase round the fields for half an hour before we realised that the sheep were either damned if they were going back or they had forgotten where their hole was. With infinite care we carefully manoeuvred them towards the buildings so that they could be penned for later removal. They were just being eased into a shed when, at precisely the wrong psychological moment, the contractor picked up a rock and prepared to relieve his frustration by hurling it at the nearest woolly moron. My scream of rage came too late to prevent him. The boulder bounced through the sheep and frightened them into scattering back through the two of us and out into the field again. When he saw the expression on my face, he turned and ran in the same direction as the sheep at such a speed that none of the rocks that I hurled could touch him.

I first ventured out into the fields with a tractor to spread fertiliser. It was one of the most satisfying tractor jobs as you could see some progress. Unlike rolling or harrowing, it did not seem to be interminable and you felt that you might actually be helping the grass to grow, which seemed to be a fairly remote prospect when all you were doing was squashing it with a roller.

Of course, there was the initial tricky job of clipping the spreader to the tractor. On our farm this was even more difficult than it had been up in Wales. The tractor still had to be precisely aligned, but no tractor that I owned ever had a properly working handbrake and there was hardly a flat surface on the farm. The tractor would be placed in position; the engine would be switched off and the machine left in gear to prevent it juggernauting off into a hedge or fence. I would then dismount to attach the implement and find that the action of putting the tractor into gear had made it lurch forward by half an inch. A quick grind of the teeth and I would re-position it again, and again, and again.

The fertiliser spreader was the only implement that was sufficiently light to allow a bit of fine adjustment if the tractor was not absolutely right and so attaching it was comparatively easy. Then six or seven hundredweights of fertiliser would be loaded into it and out into the fields to be spread. To be successful at the job, the conditions had to be just right. The principles of operation required that the granules, to make sure of an even cover, had to be thrown just far enough to fall into the wheel marks made by the tractor's previous pass. For the wheel marks to be visible you needed the grass to be either damp or long. The grass was never long because the cows ate it. If the grass was damp, it meant that it had just been raining, was raining or was about to rain again. Rain and fertiliser just do not get on. If the stuff got wet inside the spreader, it stayed in there and refused to come out. So only rarely could I find the wheel marks that were so badly needed.

On the grazing half of the farm, things were not too bad. We had split the farm into two. Over half of it the cows grazed at will, and over the other half we grew the grass that was to feed them over the winter. Where the cows grazed, the fields were small with hedges giving the cows shelter. From these, distances and direction could be calculated with some degree of accuracy. There were also the cowpats. You could peer fixedly at them as you thundered past and see whether there were prills of fertiliser already on them and adjust direction accordingly. Conclusive proof lay in a tractor-splattered pat or a flattened thistle or nettle. On the other half of the farm which was used for silage, it was a different proposition. It was dominated by one field of twenty acres and the ground was usually packed hard. Fertilising normally took place immediately after the grass had been cut in order to encourage regrowth and so there would be no chance of seeing the pattern left by the wheels.

You would get half-way across the field with your eye firmly fixed on some point at which you were aiming on the distant hedgerow, when you would realise that you had gone wrong and had no idea whether you were giving that particular point of the

53

field a double dose or had completely missed a bit further up. It was the same sort of feeling of insecurity that Columbus must have had when he was half-way across the Atlantic. That field was always four ways striped: green where I had gone over it correctly; tropically lush green where I had gone over it twice. A light and skimpy green where I had missed a pass completely, and a burnt brown where I had gone over it three times and killed the grass.

On the grazing side of the farm there were four particular hazards. The first was actually getting there. The route was through the yard and directly into the fields, which was the same route that was used by the herd. Cows, being awkward animals, do not just create a nice puddly mess if the ground is at all soft or muddy but prefer to place their hooves in the same spot as the cow that went before. By doing this, they create furrows and ridges that run at right angles to their direction of travel. This may be just fine for a cow, but for a fully loaded tractor with fifty pounds' worth of fertiliser on board it can be fairly catastrophic. Bouncing from ridge to ridge meant that the tractor underwent a series of almighty lurches with each lurch cascading a fiver's worth of fertiliser over the edge of the spreader. I could never work out whether it was better to travel at a snail's pace over this section to minimise the speed of the lurch or to take a run at it, hoping the next ridge would cancel out the jerk caused by the one previous.

The second hazard was cows. Being the grazing area, it was covered by grazing cows and they seemed to rather enjoy the fun. The essence of the job is to keep going in a straight line at a constant speed so that there is an even covering of the granules. If a cow happened to be sitting down in the path of the tractor, you just hoped like hell that she would have the sense to get out of the way in time. If not it would be a damn nuisance. Any cow within five yards of the passing tractor would be showered by fertiliser. A few took a dim view of this but most would take it as an invitation to play, and so you would grimly continue in a straight line surrounded by snorting, gambolling cows all dicing with death in front of the tractor.

The third hazard was hills. The two features that dominated the farm were the lane that cut across it in one direction and the stream that cut across it in the other. Immediately beside the stream was the tongue of woodland running alongside and up from that lay about ten acres on an almost 45-degree slope. At first

I avoided them. I did not believe that it was possible to take the tractor over them. The many rabbits on the farm would defecate at the top of these slopes and turn to watch their turds rolling happily down to the bottom where the grass grew lush and smelly. Then a neighbour casually mentioned that my predecessor travelled them quite happily so I decided to have a bash.

You have a very nasty feeling of insecurity. Point the tractor down a hill like this and touch the brakes and you will enjoy a skid that will almost certainly end in complete disaster. Point the tractor straight up and the front wheels will rise from the ground and the only form of steering will be delicate dabs on each of the rear wheel brakes. Go along sideways as was my wont and you find yourself leaning into the hill like a motor-bike rider ready to leap for safety at the first sign of trouble.

Trouble only came once or twice. The first time was when turning down the hill at the end of a pass before I learned that survivors turn *up* hills at the end of a pass. The fertiliser-filled spreader lurched round and the back wheel of the tractor rose inexorably into the air like a dog faced by a lamp-post. The adage that fear lends you wings must have been coined by somebody in this precise situation. I flew. Miraculously my feet did not become entangled in the pedals or the steering wheel and I found myself effortlessly gliding through the air and landing safely on the grass while, below me, the tractor rolled and plunged its way through the trees and into the stream. It was winched out a day later and was none the worse.

The other time was even more alarming as I had more time to think about it. Having learned the lesson about turning up hills, the tractor, in its wisdom, decided to slide backwards down with its wheels turning slowly in the opposite direction. I could not stop or touch the brakes as a final skid was sure to be the result. I was even reluctant to bail out as the machine was going so slowly that it seemed rather churlish. Using the front wheels as rudders, I managed to guide it carefully into a sapling which gave under the weight and then we plunged together back into the stream. I was forced to return to the house for a change of underpants but was otherwise none the worse.

The hills were really compounded by a fourth hazard which were the badgers. One of the greatest charms of the farm was that it contained a thriving badger sett. They had set up residence on

55

either side of the stream in a series of holes that could well have been on the farm for several hundred years. They always made me feel rather uneasy as to my territorial rights. I claimed total possession of the seventy acres to have and to hold and to do with as long as I liked or as long as my landlord should allow. Yet here were the badgers. They had had possession for centuries and would continue to hold possession for centuries after I had gone. I felt that it was just as well that they could not afford to pay for a good lawyer to explore their squatters' rights.

Where the badgers created a hazard was in their tunnelling. Their idea of a good night out was to enjoy a rollicking dig into the hillside, and the tonnage of earth and rock that they could shift in twelve hours was quite astonishing. Half an acre of the hillside was quite unusable since, apart from the difficulty of trying to steer round the sett entrances, the earth was quite likely to subside under the weight of the tractor and swallow it up. It actually happened once when I thought that I was on good safe ground. The tractor wheel went down like an express lift and a glittering arc of fertiliser granules crashed down a nearby hole. I hope it choked them.

The badgers seem to have been rarely disturbed and had therefore lost much of their normal fear of man. It was one of the stock entertainments of the farm for visitors to tiptoe out into the fields at dusk and wait for the badgers to emerge. They very rarely failed to oblige. There appeared to be two main families, one on each side of the stream. Each family had half-a-dozen holes in constant use and communication between them was over a couple of well-worn stepping stones that bridged the stream. The largest number of animals that I saw above ground at one time were two sows with two and three cubs and a couple of odd adults scrounging round them.

Purposeful viewing was not easy for as soon as they emerged they would disappear into the nettles and scrub that bordered the stream and root around for their dinner. The valley would echo with their hippo-like grunts and the frantic chittering when a couple of cubs decided to have a row. They had started off with a smooth steep hillside, but after the constant excavation of fresh burrows the hill now descended in a series of steps, the debris from each burrow jutting out like a platform from the hill and becoming hard packed by the constant passing of the animals along their paths. Three feet above ground level the scrub was impassable, but go on to your hands and knees and the whole area was networked by smooth badger motorways along which you could travel as fast as your crawl could take you.

Above one of the busiest paths there was a convenient fallen tree and on it a viewer could sit as long as he liked and watch the badgers pass beneath. Even a torch shone directly at the badgers did not seem to disturb them, and should you trickle raisins or peanuts down, you would end up with a semi-circle of badgers beneath sitting on their haunches and begging for more.

I once brought a professional naturalist to view the area in which they lived and we stumbled through the mud and nettles to view their holes and runs. The professional suddenly stopped dead in her tracks and clasped her hands together. 'Oh! How exciting!'

'What? What?' I replied, looking round for the source of the thrill.

'Look. It's a badgers' lavatory.' She was quite right too. For the record, a badger's bog looks much the same as anybody else's although the worms tend to make it heave around a bit. She found it an exceedingly thrilling sight. She even took a couple of photographs of it for her album.

If I ever ventured into the fields at night or brought the cows in early in the morning, there was always an even chance that I would encounter a badger bustling busily over the fields. They would rarely take the slightest interest in me unless I walked right up to them and then they would only look at me rather crossly before moving a foot or two further away and carrying on with what they were doing.

The dog looked on every animal from cows downwards as fair game. The cows sorted him out when they caught him in the corner of the yard and gave him the hiding of his life. The rabbits never managed to control him. His one encounter with a badger was quite instructive. He went bounding into the wood to investigate the sound of a rootling badger and there followed five seconds of growl and snap and then a sinister crunching noise and a scream from the dog. His foot was a month in plaster after that and from then on he refused to ramble the farm during the hours of darkness.

The badgers cost me money in some rather surprising ways.

Hedges were the bane of their lives. They had their well-defined paths in the wood, and the rabbits, foxes and deer had *their* runs through the hedges from which they rarely strayed. The badgers could never get the hang of going through a hedge. Their method of getting through was to charge at the thickest patch and then grunt and heave until they had managed to force their way across. Even if there was a convenient badger-sized gap beside them, they would ignore it. I once watched an animal trying to push its way between a couple of hedge roots that were about three inches in diameter. It refused to give up until I poked its behind which was familiarity beyond what it was prepared to take and it retired in a huff. All this ravaging of the hedges made them rather less than stock-proof. The badgers did not appear to be satisfied until they had bashed through a cow-sized hole, so a fair amount of time was passed in patching up their depredations.

Towards autumn, particularly after one of the droughtier summers, the grazing half of the farm would be polka-dotted with nice dry, crusty cow pats. There would be one every square yard or so. They would normally quietly break down back into the land whence they ultimately came. The cycle was interrupted one year when, in their search for bugs and beetles, the badgers excavated the ground underneath each dung pat, to a depth of about a foot. The field looked as if it had undergone a mortar attack.

Living hazards apart, the overriding problem concerning fertilising and field work was the weather. During the autumn, the cattle would come into the buildings and what machinery we had would be laid up and greased except for the tractor which enjoyed its daily yard scrape. We would batten down the hatches and endure the winter. The rain would rain and the sleet would sleet and the snow would snow, while the cows and myself would look over the fences from the yard to the fields and wait for spring to start springing and for the grass to grow.

Herein lay the problem. Every year, without fail, I would run out of silage before spring had properly sprung and was faced with the urgent necessity of turning the cows out into the fields and providing them with some grass to eat. It was one of the nastier dilemmas of the business. If the cows were turned out too soon, before the ground had dried out properly, they would churn up the fields and turn the farm into varying degrees of cocoa-coloured waste land that would not recover its full productivity until the following season. Keep them in instead and there would be a desperate period during which I would struggle to keep the milk yields up on highly expensive and poor-quality purchased feed.

The problem was compounded because in order to make the grass take off in the spring, it was necessary to spread fertiliser about ten days before the cows were actually turned out. There was a law about the first venture into the fields each year on the tractor. No matter how hard I tried, invariably the machine would dig itself deep grass-killing ruts that would not heal over until late in the summer. Even worse was the result of turning the cows out too early and that always happened as well. It would not have been too bad if they had gingerly tiptoed out into the field and delicately munched at whatever grass had been available, but to a cow the first sight of some grass after a winter spent in the yard was like showing a harem to a man who had been cast away on a desert island for a decade.

They would go berserk — staid old matrons and bright young things alike. They would gambol and kick about the farm like skittish lambs and each gambol would leave an eight-inch pothole that would curdle my blood. Coming towards the buildings from the far side of the stream was the worst, as there was an exciting hill on which they could test their skills before they hit the concrete ford at the bottom. You could see them charging along the top towards the brow of the hill like the Heavy Brigade at Balaclava. Once over the lip, they would jam on the brakes and disappear in a cloud of spray and turf as they ski-ed down the hill, each hoof peeling back the precious grass like the skin on an apple. They would finish up in an exhilarated pile at the bottom.

As messy as the cows and worse than the fertiliser spreader would be my early attempts at rolling. The purpose of rolling is to remove ruts and potholes and squash into the ground any rocks or other objects that might damage machinery later in the season. It followed that the ground had to be soft enough to be pliable but not so soft that all that was achieved was a mess. The number of revolting messes that I managed over the years was legion. Perfection was simply impossible. Some parts of the farm would be just right, while others would be still too wet. If you waited until the soggy spots dried out, then the others would be so dry that the three-ton roller would bounce from tussock to tussock like a ping-pong ball and do no good at all.

The Waterloo every time, when I would have to abandon the roller for days at a time, would be gateways. Since the cows had to cram through one narrow opening whenever they wished to move from one field to another, it followed that these spots would be more than usually muddy and would remain so throughout the year except during acute droughts. To get from one field to another with the roller, you had to take a run at the entrance. You would curve round to the far side of the field, point the tractor straight at the gate and move up to top speed with the roller clanking behind like a runaway tank. You would hit the mud in the gateway and start to lose speed rapidly. In front of the roller would build up a massive bow wave of mud. If luck was on your side, enough momentum would have been created to carry the tractor on to dry ground where the wheels could get some grip. If luck behaved as usual, the cavalcade would come to an ignominious wheel-spinning halt, and the roller would have to be abandoned, or else you could start the immensely laborious task of inching the implement out of the clinging embrace of several tons of mud.

Equally disastrous for the health of the pastures was the annual orgy of muck spreading. There are two schools of thought on the subject of muck. Muck is an expensive pain in the neck. It is smelly, clogs everything up, is produced in vast and scarcely controllable quantities and leads to water boards and local authorities persecuting the poor innocent farmer when he tries to dispose of it. The other school believes that it is a valuable fertiliser that should be cosseted, cherished and treated with all the respect that you can muster and should be carefully smeared over the farm at the most opportune time· to encourage crop growth.

My own opinion of farmyard manure fell firmly into the first category. The second may or may not be valid. All those who sold shiny metal containers in which to put the stuff certainly tried to persuade me that it was so, but one of my prime business motivations was to avoid spending any money unless a quick and reasonable return could result and, in spite of all the sales talk with which I was inundated, I have yet to hear of a farmer who was made rich by farmyard manure.

My predecessors had obviously been of the same opinion as there were several derelict methods of manure disposal on the farm. The first of these had a certain elegant simplicity. The evidence lay in a grassy tennis court that lay in a dip beside the stream. In my first week on the farm I investigated this curious stretch of ground which was in an acre or so of wilderness just below the milking parlour. I threw a questing stone into the middle of it and was alarmed to see the entire surface of this lawn heave and undulate as if it was manifesting an earthquake. Obviously there was fluid of some kind which could have been a potential cow trap that should be removed. I carefully dug a little ditch the six feet from the stream to this pit. What came down the little ditch was straight out of the sewers of hell. A strange black liquid with the appearance and consistency of treacle with a smell the memory of which still has the power to awake me screaming from a deep sleep. A few pints of this substance crept down to the stream and immediately fish throughout the interconnected river systems of southern England floated belly up on the water. I hurriedly, gagging, filled in the ditch as the head of the spade turned white and crumbled to dust in my hands.

This was the relic of the earliest method of dung disposal on the farm. A channel had been dug straight from the buildings to this pit, and when the stream was in flood a sluice gate had been opened at either end so that the stream had rampaged merrily through and had swept the whole lot down into the English Channel. These days the authorities frown on this beautiful system, and although it is still practised in some rather surprising places, it can cost five hundred pounds every time that the water board catches the perpetrator at it.

The second system at a slightly higher level on the archaeology of the farm's past was the pump and pipe method. This system had its relics all over the place. At its heart was a large, concrete-lined pit into which all the goodies had trickled. From there a pump, so the theory went, would send the slurry along hundreds of yards of aluminium pipes to the nethermost regions of the farm where it would be harmlessly sprayed over the fields.

Where the theory and the practice fell out was that the pit was situated at the lowest point of the farm and no pump could be found that was pump enough to shove the slurry up the couple of hundred feet or so that was needed. The farm was littered with these aluminium pipes with the rigidity of macaroni. Try as I might, I could not get rid of them. The scrap man would refuse to touch them, and the only use that I could think for them was as fence and hedge pluggers. There they would look most formidable until a mouse decided to scuttle along them when they would sag tiredly to the ground.

On my take-over of the farm, the slurry disposal was in some disarray, the pump system having just been abandoned. A ramp was in the process of construction, up which the slurry was to have been pushed over the lip into a spreader and thence straight out into the fields behind a tractor, I was not all that keen on this method for several reasons. I was too mean to invest in a spreader and a tractor powerful enough to pull it. Most of the dung was produced, and had to be disposed of, in winter and I was only too aware of the effects that the tractor would have on the fields. I had also just heard the saga of a cow that had decided to jump into one of these spreaders. It may sound a pretty daft thing for a cow to do, but no cow has ever been deterred from any action merely because it may have been a little daft. Once the animal was in there it was almost impossible to get out. Every time that it struggled to its feet, it would slip on the curved and shitty bottom and subside again. The whole implement had to be packed with straw, and blocks and tackles attached to the cow before it was eventually winched out.

I became a pit man. This has a simplicity that is almost on a par of the stream method and has the advantage of being rather more benignly regarded by the authorities. Devotees of pits dig a dirty great hole and shove everything into it and have a gargantuan spread once a year when the farm is abuzz with crawlers, spreaders and contractors. The creator of the pit owned an enormous crawler with a bucket on the end of a long arm. The years of practice allowed him to use his bucket to scratch his nose and unwrap his sandwiches. He came initially with his entourage of spreaders to remove the winter's droppings left by the herd of rat-like bullocks that were infesting the farm when we took it over. Their owner had obviously been a convinced pitter although he had neglected to dig himself a pit. Muck was lying about all over the farm, much of it liberally reinforced with the acres of plastic sheet that had covered the silage and bound with the twine that had held his straw bales together.

The machine filled up a spreader with its first load and it bumbled off behind a tractor to be dumped. The tractor driver was supposed to slip the spreader into gear to revolve the chains that chucked the slurry out in nice bite-sized chunks and return for another load. He returned because the polythene had wound its way round the shaft of the spreader and prevented the chains from chucking out the muck.

We both looked into the machine's interior. It was dark and full, redolent of the smells of six months of dung. Somewhere inside, the polythene was waiting to be extracted from under a foot or so of dung. The contractor looked at me and I looked at the contractor.

'How are you supposed to get it out?', I asked.

'You have to get into the spreader and feel about under the surface of the slurry and then cut away the polythene bit by bit.' I digested that one for a bit. 'That won't be much fun.'

'No, it won't be,' he replied. We looked hopefully at each other. Neither of us made a move.

'I'll help you get into the spreader,' I said encouragingly. He looked wildly round him for inspiration. He clutched frantically in the area of his kidney.

'It's my back you know. The doctor says that I shouldn't do any bending.'

'Can't you sort of empty the dung out of the spreader by asking the bloke with the bucket to turn it over?'

'Oh, I couldn't do that. It might damage the machine.' I had a horrible feeling that I could see the direction in which the conversation was leading.

'Of course, I don't know the layout of the machine under the slurry and I could well damage it myself if I went rooting about in there.' He brightened up. 'Don't you worry, I'm quite willing to take the risk.'

'Thanks a bunch,' I replied and clambered reluctantly in. I felt nervously under the slurry with my gumboot for the clogged up central shaft that ran the length of the implement. The contents had the consistency of heavy mud. My feet slipped off either side of the shaft. To save myself from squelching my vitals on the thing, I threw myself forwards and ended up full length on top of the muck. I looked over the edge of the machine at the contractor, the slurry dripping from my hair.

'Could have been nasty, that,' he said.

'Yes, wasn't I lucky?'

'You were lucky all right.' He handed me a pocket knife and moved upwind while I plunged an arm into the chilly interior and hacked off a piece of polythene. It took ten minutes to clear the shaft before I was able to climb out. Another spreader entered the yard.

'Here', said its driver. 'Since you stink like a zombie's shit house already, you won't mind clearing some plastic out of mine.'

Once the muck was cleared and spread all over the farm, some of its drawbacks as fertiliser showed themselves. It made the grass underneath it useless until the rain had washed it in, and woe betide you if it was spread before a dry spell as it would stay there for months. Sure enough the grass grew round it and through it and looked lush and green and long, but that was hardly surprising as no cow would touch it until all the nasty taste had been washed off.

One aimed to produce nice, clean, pure piles of slurry if there is no basic contradiction there, but the stuff had a magnetic attraction for old plastic bags, syringes, retired gumboots and hunks of wood that invariably ended up on the pastures of the farm. All of these had to be laboriously cleared up after spreading. I once collected a rather nice cigarette lighter, dropped by some careless cow or other and it was little the worse for six months in the dung.

Once we had scraped all the extant slurry up and spread it, the contractors went home and left the place in charge of the hole artist and a mighty hole he dug. It was about twelve feet deep, lined with clay and shale, and looked large enough to contain the outpourings of half the cows in the country. There was even a Sign of Approval. As he switched off his machine on completing the dig, a sunbeam slashed its way through the cloud covering and lit up the pit. As we goggled at the ludicrous inappropriateness of the Sign, it was compounded by a real live white dove that came fluttering over the trees and settled in the bottom of the pit, lit by its sunbeam. The dove tipped the performance over the edge into pure schmaltz which was only redeemed by the bird mislanding in a puddle, so that I had to clamber after it and rescue it from drowning. It dried off rather ungraciously before returning to the local pub whence it had come.

In general, the slurry pit was a great success. Another of those curious agricultural laws applied. The pit would always appear to be full except for two weeks after it had been emptied. In spite of this appearance, you could go on scraping the farm effluvia into it from November to March and it never overflowed. This may have been helped by the skill of the constructor, who built in a couple of lovely little drains that stealthily tapped off much of the liquid straight into the stream. Nevertheless month by month, the solids piled up high above the lip of the pit, but the walls stayed intact and there was never an avalanche.

Each year, at spreading time, the digger would occupy its duller moments by enlarging the pit, but the same inexorable law of fulfilment would apply the following winter. The spreading was always pretty ruinous on the fields. The art of husbandry lies in timeliness. At 2.33 pm on a certain afternoon the conditions might be just right for some particular field operation. Such precision was never possible with muck spreading. The spreader contractor would promise to liaise with the digger on the farm during a certain week and, if you were lucky, they would turn up a month either side of the right day.

Another agricultural art that I never really mastered was that of building a working fence from scratch. Round the slurry pit, as you might imagine, this failing was rather noticeable. Picture a pit with a nice smooth top, frequently covered in greenery as the summer wears on and picture underneath the greenery a twelve-foot depth of quicksand-like slurry. It was obviously in my own interest to do a really thorough fence job round such a lethally dangerous spot.

This was my one major attempt at a free-standing fence and I worked at it in Forth Bridge painting fashion for several years. It had to be a mixture of barbed wire and netting to keep out children, cows and Safety Inspectors, but it never really kept out any of them. The principle, I am sure was right. Hammer the posts in good and hard. That was not all that easy for a start as the walls of the pit were soft, having been recently disturbed, and a tight, firm post was, I think, an impossibility. Then I would string along the top wire and tighten it so that it twanged high 'E' when struck. Next I would move down and tighten the second wire which meant that the top wire would become slack. So I would move back and retighten it which in turn slackened the second wire. And so it went on. There has got to be a way of doing it right, but I never found it.

This fence not being quite as stock proof as it might be meant that some of the irresponsible cows would jump it and go for a mucky swim. The first time that I realised this had happened was in the parlour one morning when 45, the bull's beloved, came in plastered with slurry from top to toe with only her head and neck still clean. She had fallen in and scrambled out by herself. I gave her an extra pull of concentrates for gallantry since she had saved me the £500 that it would have cost to replace her had she drowned. Usually there was a good stout crust on top which was the saving grace of the pit as well as its biggest deceiver. The goodies in the slurry that floated would rise above the urine and gradually dry out to create a layer on top that might or might not support weight. If it did, fine, you could roast oxen on the top with no trouble. If it did not, as on the frozen Thames, the first you would know about it would be when the muck swallowed you up — and I can think of better deaths.

The spreading contractor once told me that he had emptied out a pit and had found the remains of three cows at the bottom. When he gently broke the news to the farmer, his reaction was a snap of the fingers and 'So that's where the blighters got to. I thought there were one or two missing.' I never lost a cow in the slurry pit. Once every two or three months, one would decide to dice with death but the crust would support her weight once the four hooves had broken through and, with a bit of persuasion, she would undulate back to dry land rather like a sea lion. In the heat of excitement at the prospect of losing a cow, I would find myself standing in the middle of the pit belabouring her back end to encourage her exit. I would stop in a cold sweat when I realised what I was doing and gingerly return to dry land. Latterly I created a couple of snowshoe-type attachments which slipped over my boots and, with these afoot, I roamed the surface of the pit with confidence.

The difficulties with the fence were not helped by the digger which came emptying each year. Its operator had to rip down my fence to get at the pit and it seemed to be re-erected slightly tattier each time. It is a mistake to hire a digger whose operator is a perfectionist, since dung clearance is one of the few jobs where it **can be a positive disadvantage.** The operator believed that it was a slur on his professionalism if there was the least trace of slurry left in the pit after he had cleared it. This meant that the bottom inch or so was scraped up with the addition of a couple of inches of rock, shale and rubble and this duly clattered into the spreaders and ended up on the fields along with everything else.

In spite of certain drawbacks, the system worked very well. The slurry only cost money for one day each year when it was emptied. Otherwise it was just shoved into its hole and forgotten. That was the sort of farming I was after. Everything that did not lead to yielding money, I wanted to be able to ignore. I did not possess sufficient expertise to fritter it away on inessentials.

# Chapter Five

One of the less attractive aspects of keeping cows was their insistence on producing calves. This was an entirely voluntary act on their part. All I did was stick in Joe and let nature take its course. Any cow that felt tired or exploited was perfectly entitled to refuse to conceive for a bit and give herself time to recover but very few of them ever availed themselves of the opportunity. Infertility is the most common cause of a cow being retired to the knacker, but during the five lactations that the majority of the cows had on the farm, not one of them was ever sent to heaven on those grounds.

The result of using Joe was somewhat disconcerting from the point of view of management. In my studies of various agricultural tomes and through analysis of advice, it was apparent that the vital ingredient for financial success was that the cows should calve every year. The calving index towards which you should strive was the magical 365 days, and you were supposed to slave night and day to keep it down to that figure and prevent it creeping towards the more common 400 days. The herd obviously had failed to read the right textbooks since they, over their five lactations, preferred a calving interval of 330 days. Although this suited me fine from the point of view of producing milk, all my studying had been towards the management of the autumn-calving cow. Since the animals were jumping back a month each year, it meant that I had to do some very swift mugging up on the management of the summer-calving cow and, latterly, the spring-calving cow which calls for very different methods of approach.

The greatest exponent of the short calving interval was 45. She was the animal that Joe relied upon for his daily intake of milk so I suppose that she thought she ought to contribute to the costs of keeping her in some way or other. Over five calvings, her average interval was nine-and-a-half months which, when you consider that the cow's reproductive cycle is almost exactly the same as the human, is not bad going. She even produced three sets of twins. Her finest hour was when she decided to drop a live calf eight months after the production of her previous set of twins.

When she started to show signs of calving, I could not really believe the evidence of my eyes. Cows are not supposed to reproduce at quite such a speed and so the vet was called in to investigate and try to establish what she was up to. I manoeuvred her into the bull pen, which was where all medical examinations took place, since Joe would have been decidely put out had it been put to its proper use. The vet, when he arrived, turned out to be young and not very experienced. He, too, was a little baffled at what 45 was up to, so he stuck on a long polythene gloved and chased up her back end.

Number 45 took rather unkindly to this and pushed his arm out again and kept on pushing to emit an absurd little calf, obviously premature, about the size of a poodle. Astonishingly the animal was alive if very weak. Harsh economics at the time would have directed that the animal should have been left to die since the calf price was on the floor and the value of a calf such as this would have been somewhere down in the basement, particularly since the vet had been clocking up a pound note every couple of minutes since he had arrived on the farm.

If my job was keeping animals, then my job surely included keeping them alive and so we both went to work on the calf.

'We'd better get some colostrum into the thing,' said the vet. If a calf does not suck some colostrum from its mother in the first couple of hours of its life, it is very likely to fall victim to the multitude of infections to which its flesh is heir. I hauled the calf to its mother's udder. The calf lay flat out, not taking the remotest interest.

'I'll try shoving a tit into its mount,' I said.

'That's a good idea.' I tried it and there was no response. I looked optimistically at the vet for suggestions. 'Er, I know. I'll go and get a tube and you milk some colostrum from the cow and we'll pour it directly into the calf's stomach.' Brilliant, I thought. He disappeared while I sat down on the straw and extracted some of the sticky yellow liquid into a bucket. The cow was too busy trying to lick a bit of enthusiasm into her calf which was still lying half dead on the floor to worry about me scrabbling about round her udder. The vet bustled back.

'Right then, you hold its head.' I held its head as the vet carefully fed the tube into the calf's mouth, pausing to listen at it to make sure that it went down into the stomach rather than into the lungs. He fitted a funnel onto the exposed end of the tube.

'OK. Now pour some of the colostrum down.'

'Right. How much should I put down?' The vet looked a bit disconcerted.

' I'm not very sure. I've never done this before. Do you have any idea?'

'No,' I replied.

'Oh well. Not to worry. How about a good slosh.' I poured down a good slosh. Nothing very much appeared to be happening. 'And another slosh.' I sloshed. 'I should think that's about enough,' he said.

'There's only a little left. Shall I finish it up?'

'I suppose you might as well.' I poured the last bit in and the vet pulled out his tube. We sat back and looked at the calf. Its condition had definitely changed.

'Do you think its belly is meant to be swollen like that?' I asked. The beast looked as if it had swallowed a football.

'Hmm. I think we may have given it a bit too much,' said the vet thoughtfully. The beast was staring with distended eyes giving out the most appalling grunts and threshing its legs feebly at nothing.

'I could have made a bit of an error here. Hang on and I'll go and check on the radio and find out how much colostrum we should have given it.' He disappeared to consult with his colleagues and I gloomily watched the calf. He came back.

'How much did we give it?' he asked.

'I didn't get much out of the mother. It can't have been more than five or six pints.'

'Hmm'.

'Well? How much should we have given it?'

'They think that between half a pint and a pint would have been ample.' We both gloomily watched the calf. It appeared to be **choking**.

'Shall we try and syphon some out again?'

'I thought of that. But even if we could it is probably too late,' he said. 'It's really quite an interesting situation.' I did not really agree. 'It seems a very expensive and laborious way to kill a calf.' He brightened a bit. 'Oh, it's not dead yet. You never know.'

I put the beast in a straw-filled box under an infra-red light for a couple of days. Its grunts slowly became less painful and in a series of monumental shits, its stomach slowly subsided to normal. It survived this intestinal violation and hung around the yard for a month or so drinking a gallon of milk a day before I bounced it into the local market where it fetched seven pounds, having grown to the size of a Labrador.

Calves are the cross that the dairy farmer has to bear from before their birth to the day that they are sold. They are a necessary evil and had the power to dredge from my soul all the emotions that tie man to his animal ancestry. My first batch of thirty-five heifers elected almost unanimously to present their calves backwards. In my pursuit of letting the cows do what came naturally, I never insisted that they came into a germy pokey shed to bear their offspring but allowed them to select their own time and place.

I suppose that the normal progression of a bovine delivery is the commencement of labour followed by the bursting of the water bag, with the calf being born in the attitude of high-board diver within an hour. These backward-presenting heifers took considerably longer. The pressure of the calf's head in the pelvis is, I believe, supposed to stimulate the cow to heave and eject her calf. With the calf coming backwards, there is no such stimulus and so the whole process comes to a grinding halt. At one point half the animals on the farm seemed to be going about their daily business with a pair of back legs sticking casually out of their behinds. The local opinion was that these reverse deliveries were caused by bringing the animals to the farm on a lorry too close to their confinement.

In many cases it would be my job to grab hold of the projecting legs and give a helping heave. When a heifer lay down to do a bit of concentrated delivering under a convenient hedgerow, I would stealthily creep up behind her and try to affix my calving ropes, or more accurately, my baler twine. Usually the mother would turn and cast me a decidedly supercilious look before rising to her feet

71

with dignity and retiring to another hedgerow. I would stalk my quarry round the farm and only rarely manage to hook on. It is one of the disadvantages of single-handed farming that it can be very difficult to corner or herd animals if you are on your own with only a mentally deficient dog for company. Frequently and expensively the vet would have to be called in to help with the capture and extraction of the calf.

Most of the animals decided that the farm bog was the perfect maternity area. This was down by the stream at the edge of the badgers' home territory. It was, of course, the furthest point on the farm from the buildings and so the midnight maternity check would involve a long, cold and soggy walk. In the centre of the bog on an island lay a thicket of gorse, and the prospective mothers would force their way through to the middle of this before lying down to give birth. It was the safest place on the farm but a bloody nuisance. Stupider cows would select a dry patch on the edge of the bog by the stream to push out their calves. What they failed to take into account was a six feet drop off the edge of this platform into the water, and many a time it was necessary to fish out a sodden, half-drowned calf.

One particular delivery sticks in my mind. The mother was right in the middle of this gorse with her calf coming out backwards and apparently firmly stuck. I tied the ropes on to its legs and knotted them behind my back so that I could get down and really heave. This method can have unfortunate repercussions if the cow decides to take off as you can find yourself bouncing happily along in her wake like the typical shot cowboy with his foot stuck in the stirrup. This cow stayed put. I would heave and she would heave in sympathy. I looked behind at one point to make sure that I would not land in the middle of a gorse bush should the calf suddenly pop out. There, sitting about eight feet away was a fox waiting patiently for the afterbirth so that it could have its meal.

The cow was busy doing her job as best she could. I was busy doing mine as best I could, and the fox was waiting to do its job as best it could. When the calf eventually arrived, the fox retired a couple of feet and waited until the excitement was over and we had moved away to the buildings before unconcernedly moving in to clear up.

Inevitably it is those calvings which give trouble that stick in the mind. Usually I would not be around for the normal deliveries, and the first I would know of them would be the sight of a healthy little calf in the corner of a field with the mother standing guard over it. In this first batch of calvings, I only lost one in spite of most of them being born backwards. The one was son of 13, a large, amiable and totally moronic cow. who produced her first offspring with no difficulty at all and then sat on it and squashed it flat.

I gradually became more skilled in the arts of bovine midwifery and grew to depend less upon the vet. If a cow had any difficulties in pushing out her calf, I would shove in an arm to investigate. It is a most peculiar sensation akin to pushing the arm into a lovely warm bowl of tripe. I then had to grope around desperately for a nose, leg or some recognisable part of a calf's anatomy, sort it out and pull. Every time that the cow contracts, the blood is squeezed out of the inserted arm and I am sure that a prolonged squeeze could lead to gangrene.

Calves were presented in various ways. There was the normal diving approach. Sometimes there would be just a front or a back foot offered and then I would chase up the back end of the cow to disentangle the mix-up. One cow, every time, would push out the calf's head and nothing else. One time I spent an exhausting twenty minutes chasing her round the field with the calf eyeing me sardonically and even blinking at me occasionally. The sheer physical strength required to extract some reluctant calves was enormous. There are devices on the market to help. There is a sinister corkscrew-type thing which is placed against the cow's behind and is liable to pull the calf to bits unless care is taken. There is even a school of thought which believes in trapping the cow's head, attaching ropes to the calf and then to a tractor. This has the slight disadvantage of occasionally pulling off the cow's head. The vet used a block and tackle which was too complicated for my brain to understand. So I used a gate, tying the rope on near the hinges and pushing the other end, thus providing enough leverage for all but the most bashful.

One of the most frustrating parts of the business was when a full-term calf was born dead. Not infrequently, the umbilical could break or be pinched during delivery and lead to the death of the calf. This is always infuriating. On receipt of a calf, the duties of the midhusband are to ensure that it is breathing and that the windpipe is clear. Sometimes a calf would be born with a good strong pulse hammering away at the base of the umbilical cord and would just refuse to breathe. Then I would leap into swift action: rub the animal's flanks with straw; chuck cold water over it, and even swing it round and round my head which is quite a knack with a 90 lb calf. Sometimes this effort would be rewarded by a convulsive heave of the flank and the calf would deign to live, but all too often not.

One of the last calvings on the farm was one of the most annoying. The cow buckled down to it beautifully, but after an hour or two of hard work all she had to show for it was a ratty little tail poking out. I had a desultory scrabble around her interior but could not work out which was which and what what. I summoned the vet and we tramped our way over to the gorse patch and he got to work. He sweated away for fifteen minutes before pulling out a huge calf with a strong pulse and no respiration. All the usual tricks were tried including the ultimate weapon, mouth-to-mouth resuscitation, and if you wonder why that is tried last rather than first, just you try it on a gungy newly born calf still steaming from its mother's uterus. In spite of all effort, the beast was dead and the vet stuck his arm back up the cow to see if he had done her any damage during the delivery. She had another calf still inside and exactly the same procedure with the same result ensued.

Once the calf had been born and, in all but a couple of cases each year, born alive, the next step was to bring it into the buildings. This would be the beginning of a short but miserable relationship between it and me. The purpose of the agricultural exercise in which I indulged was the production of milk. The production of calves was very much a secondary object. The lactation curves of the cow builds up to its peak a month or two after the birth of her calf and dwindles away in a year to virtually nothing. The calf, then, is a necessary evil in the system.

First of all I had to ensure that the calf was left with its mother in the field for the optimum length of time. If it was left too long the mother would develop a severe case of mother love and, worse, the calf would take one look at me and run like the clappers when I approached it. Left too short a time and the calf would not have had time for that all-important first feed of colostrum and, again worse, would be too young and dozy to walk; carrying calves leads to slipped discs and heartbreak.

Even when your calf has sufficient clue to get up and walk, it is far from plain sailing. There is no guarantee that it will move at your request in the direction that you desire. I must have brought in between two and three hundred calves but every time I was faced with an unknown quantity. The rare and wonderful exceptions were the few that trotted happily down the field by the side of their mothers. The normal animal did not and no method that I came up with would enable me to drive or lead a reluctant calf with any degree of efficiency. Thump it on the bottom and it would stop dead. Pull it forwards and it would try to go backwards. Get cunning and pull it backwards in the hope that it would go forwards and it stopped dead again. One semi-viable method meant that you had to pick up its back legs and move forward as in a child's wheelbarrow race. Another method was to turn it over on its back so that its feet would not get tangled up and drag it along. The best method was to wait by the side of the farm lane until a rep came visiting and hitch a lift down to the buildings on his bonnet with the calf clutched firmly in your arms, praying that he did not touch his brakes too hard.

Calves could quite easily damage themselves during this initial stage of their lives. The runners could give themselves real problems. The runner was the calf left frationally too long with its mother in the field so that it had had time to sort out the world in front of its nose to the extent that it knew that I was no part of it and should therefore be avoided at all costs.

One calf leapt to its feet at my approach and ran straight into a tree, knocked itself out, turned blue and stopped breathing. I spent ten minutes giving the brute mouth-to-mouth resuscitation trying to avoid the flashing tongue of its mother which was licking me and it indiscriminately before, with much snoring and scrabbling of the feet it took over the chore of keeping itself alive. Another calf ran straight into a hedge and skewered its eyeball on a thorn and blinded itself in that eye. The cow and I had the sympathetic horrors and had to sit down for a bit. Another did a high dive into the deepest pool in the stream and, much to my disgust, I had to follow suit to fish it out before it drowned.

Once the calves were safely locked in the bull pen, there were two main problems. The first was excessive mother love. Some cows were worse, or perhaps it should be better, than others. The brutal reality of dairy farming meant that the calf never again sucked from its mother. Her job was to fill up my milk tank. If the calf had been allowed to continue to suckle, it would have led to udder problems with the cows. The Friesian is bred to give six or seven gallons of milk a day. She will only give this milk properly when stimulated to release the hormone oxytocin by either the milking machine or the action of the calf sucking. The calf that could cope with seven gallons of milk a day has yet to be born — even Joe in his prime could not manage that much. This would mean that the cow would still be brim-full of milk after the calf had had its suck but would have used up her supply of the hormone and could not release the rest of the milk properly in the parlour. Congested udders led to the dread disease of mastitis.

All this led to deprived motherhood and she would stand outside the bull pen and bellow until the pangs subsided. The pangs lasted from two hours to two months depending on the individual. Sometimes the cow would fix on me as a likely calf substitute, probably because she could smell her calf on me. The most unlikely and unfriendly cows would dog my footsteps and breathe warmly and lovingly in my face or rasp away at me with their tongues or even dangle an udder at me in the off chance that I might fancy a suck. Deprived motherhood and incipient motherhood were not compatible. 'Deprived' were only too likely to try to take over the calf from 'incipient' whenever it was produced, and vast and bloody fights for possession would ensue with the calf being trampled in the middle.

I soon learned to segregate the cows that were expecting the blessed event from the others. Usually I would stop milking them eight weeks before they were due and give them a bit of a holiday away from the milking parlour. Before I became wise to all the problems, I erected only a single strand electric fence between the dry cows and the wet cows, many of whom would have been recently deprived of their calves. One dry cow managed to calve with a mighty heave and squirted her offspring over the fence into the field with the wets. Alerted by the noise of combat, I went to investigate and found a rugby scrum made up of thirty cows all kicking, butting and bawling with Joe looking on disapprovingly. I had to wade into the middle of this maelstrom and rescue the calf that appeared to be in imminent danger of being turned into instant steak tartare. Walking back down that field was the only time that I was really frightened of the cows as they all fought to regain possession of it from me. When mass hysteria grips a herd, it is an awe-inspiring experience to walk through it.

Once the calf had been wrested from its mother, it was time to begin that most dread of rites — bucket training. To train a calf to take its mother's milk from a bucket rather than straight from the teat is a true test of patience and sainthood. You need to take the process in steps. First, half fill the bucket with warm milk from the calf's mother and approach the little dear. You will find that the calf will run away from you. Corner your calf and show it the bucket: the calf will bellow, and its mother will come and rampage up and down outside the pen. Drive off the mother to prevent distractions to the calf and corner it once more. Dip your finger into the bucket and stuff it into the calf's mouth. The calf will then grind its teeth and take much of the skin off your finger. At this point you will feel the first temptation to belt the calf. Resist it. Redip your finger if you can extract it from the calf's mouth and repeat the process. Eventually you should begin to feel the calf suck.

Next slowly try to lead the sucking calf down towards the milk in the bucket. It will withdraw about three inches before you reach the desired spot. Redip your finger, never mind the blood dripping from it, and repeat the process. The calf will withdraw its head a foot before the milk. Repeat the process again . . . and again . . . and again. Finally the calf may actually touch the surface of the milk with its nose before snatching it back. Resist the temptation to force the animal's head into the bucket. Remember that nature has programmed the little horror to suck from an udder above its head and not a bucket.

Redip the fingers and lead the calf back towards the bucket. Remove the calf's foot from the bucket and all the odd bits of straw and dung that came in with it. Redip your fingers. Do not belt the calf merely because it has upset the bucket. The milk will have become cold by this time and would have given the calf indigestion had you persuaded it to drink any.

Return to the dairy and collect some more milk from the bulk tank. Spend ten minutes warming it up again because the tank is refrigerated. Return to the calf. Corner it once more. Dip your finger and recommence the operation. Aha. This time you manage to get the calf right down into the milk. It puts its head in and starts to drown. Yes, you should feel more sorry for it. So the process goes on. Eventually the stage will be reached when the calf has learned that the milk is at the bottom of the bucket and then you can try to withdraw your fingers. You will then discover that the calf is firmly convinced that the fingers are a vital catalyst, without which the milk cannot be consumed, and another long process gets under way when you try to disabuse the calf of this idea. It is a process that demands all the patience you can muster and, when you run out, you will find that you have not even got as far as convincing the calf that the milk is in the bucket.

The more calves that there are to bucket train at any one time, the more patience you will find you have, mainly because the karate hand will have become too tender to remonstrate with the animals any further. A neighbour of mine, who is closer to sainthood than most, had a calf in a rather nice step-over leg lock and was about to forcibly submerge it in the milk until it either drank or drowned when the calf let out a bellow. Through a block wall and two shut gates charged its mother in response and skidded to a halt, breathing heavily and pawing the ground, a couple of inches away. Neighbour gave the cow a weak smile and carefully disentangled himself from the calf and was forced to try again more gently.

That same neighbour was asked to come to a farm owned by a sweet old lady who was having a bit of trouble with her calf feeding. He went to look at a pen of twenty-five excellent looking calves who came over to the gate to greet him and suck his fingers. The old lady put her head round the edge of the building. A snort from the calves and they piled themselves up in a corner and looked at her with rolling, terror-stricken eyes. Neighbour looked accusingly at the old lady who hung her head and sloped guiltily away. The calves disentangled themselves and came back to the gate.

Some people can do it. 'Do you find it difficult to bucket train calves?

'Oh no. We find it very easy.'

'Really?'

'Yes. Well of course we keep Jerseys rather than Friesians and I think they must be a bit easier.'

'Lucky you.'

'Yes. Of course the trouble with Jerseys is that the bull calves are worthless. They're no good at all for beef.'

'What do you do with them then?'

'Oh, we put them in the deep freeze . . . it's rather distressing really. I have to go out and shoot them and they will insist on sucking the end of the gun barrel.'

Joe, being father of the calves, did not produce purebred Friesian calves. His offspring were all Hereford crosses and designed ultimately to be turned into prime beef. This was a disadvantage in one way because it meant that I could not rear them to become herd replacements. It was an advantage beyond rubies because it meant that I could get rid of the little horrors as soon as was decent.

I believe there is a law as to what is decent but I could never find out what it states. Our period of decency was retaining the animals on the farm until the umbilical cord had dried up and then they were bounced into the nearest market. Selling calves is a strange business. The first crop that was born on the farm was sold when the calf market was down. When the calf market goes down, it goes a long way down. We still have the cheque for the first calf that we sold amounting to £1.63, less 15 per cent from the auctioneer, tolls and 50p from the haulier who took it to market for us. If you consider that the calf would have drunk several gallons of milk, it is possible to understand why some farmers thought that the appropriate greeting for a new-born calf was a sharp blow on the skull.

My second venture to market was with a bunch of eight calves and the cheque for them amounted to £13. I became a bit cross at that point and complained to the auctioneer, convinced that the calves were being ringed when the buyers realised that they were faced by a vendor still very wet behind the ears. The week after that I sent two calves and got back a cheque for £20, which was riches indeed. The fluctuations in the calf market made any form of budgeting impossible. The year's crop of calves could be worth £60 one year and, the very next, £5,500. Hardly a dependable and reliable source of income.

I managed one totally trouble-free calving season. One of my neighbours fancied a bit of beef production and, as Joe was slabbing on the flesh thanks to his milk diet, he decided that if he could buy a bunch of calves that would end up by looking like Joe, he would not go far wrong. This was bliss. Every calf that was born on the farm that year became his property the moment it saw the light of day. My sole responsibility was providing the odd bucket of colostrum so that his family could do the feeding. We worked out a fixed price based on what calves were averaging in the markets and each animal, regardless of sex or size, went for the same amount. It was the nearest I ever got to the ideal towards which I strived. The ideal was that each cow should lie down by a hedge, heave mightily and eject a smart pigskin wallet containing a bunch of crisp new ten-pound notes.

Normally, I would have to flog the smelly little brutes round the markets. There were two ways of getting them to the market. The first meant calling in the haulier. He charged money for his services and always called at the least convenient time — either during milking, which I would have to interrupt to concentrate on shifting calves and get kicked by impatient cows when I returned, or else after milking when I would be perched on the loo with the *Sun* and would have to pick up everything and run. The haulier could only really be justified if the farm was suffering from a severe infestation of calves at the peak periods. Having a bull of the potency of Joe meant that there could be a fairly rapid build-up of calves. His record was fathering fourteen calves in eight days.

More usually, there would be just one or two calves to go and these would be stuffed into the back of the estate car and winged to market at about 80 mph. Speed was of the essence because calves smell. The animals produce dung of the most astonishing variety of textures, colours and smells during the first few days of their lives. The first effort was invariably ejected as I would be carrying them in from the fields. This was always green, long, not too smelly and would smear itself all over my front and hands and had the sticking capacity of Bostick. Once the animal had rid itself of that little load, its main armament was smell. Cow, one quickly became used to, but the smell of calf dung was totally different and comparable only to cat, of hallowed memory. For the few days before they were sold, the calves would practise building up the odour bank, the colour running the gamut through yellow, orange, white and grey depending on diet and inclination.

The beasts would achieve their noisome climax during the trip in to market which explained the need to keep the speed of the journey as close to eighty as possible. In spite of having the doggy guard in place to keep them in the back and covering the surfaces with sheets of plastic, the yellow turds would sneak into inaccessible places and their smell linger in the air for days.

I only once broke my rules and went out and bought some calves as opposed to merely selling them. This was done in order to overcome Joe's inability to produce nice little Friesian heifers for me. I invested in five of them and palmed them off on a neighbour to rear. The contract stated that he would do all the work in exchange for nothing more than money and would provide me with five nice little heifers on the point of calving two years later. I thought it might be cheaper than buying them as adults or something.

It was an admirable idea in principle but, to my dismay, the neighbour decided to expand his own dairy herd and needed the space that these heifers were occupying. He dumped these five little horrors back on me and, willy nilly, I was forced into the art of heifer rearing. Fortunately they conveniently slotted into a two-acre field right in front of the farmhouse which was rather remote from the rest of the fields and a bit of a nuisance to manage. They came back early on in the year when they were about six or seven months old and went straight into this field before the grass had really started to grow, and so they received a bucketful of nuts or sugar beet pulp in order to keep them growing, with one annoying result. Their field ran down the edge of the farm lane between the buildings and the house, and every time that anyone walked up or down the lane, the heifers would keep pace with them, bellowing and bawling for food. It became a little wearing on the nerves after a few months and I took to creeping along the side of the hedge bent double in order to avoid attracting their attention. I kept this up until I was surprised crawling down the lane on my hands and knees by a visiting rep who found my embarrassed explanations rather unconvincing.

The heifers were actually rather endearing. I found that the easiest way that a cow could worm herself into my affections was if she appeared to be pleased to see me. Normally a cow is about as emotional and phlegmatic as the Albert Hall and her interests, apart from food and sex, are solely concerned with doing as little as possible. These heifers, no doubt because of their youth, would come gambolling up for a scratch or a play if anybody went near them. They and the dog gave each other endless amusement. The dog would go into the field and lie down. The heifers would run up to within a couple of yards and then hesitantly sniff their way forwards. Just when they were about to lick his coat, he would leap to his feet and be chased a couple of circuits round the field before he would lie down again and be stalked once more. They kept this game up for hours at a stretch.

It is always essential when you handle stock that the beasts should know who is boss and show the proper respect that is due to Man as the superior species. If you have ever seen a couple of cows having a brisk altercation as to which of them was higher up in the hierarchy, you will have realised that man would stand very little chance if they decided to treat him as an equal. These heifers did not know enough to show the proper respect and, flattering as their attentions could be, they could also be rather overwhelming. Five hundred pounds of heifer leaning against you and scratching herself against your leg could soon knock you flat. It became even worse a few months later when they reached puberty and the joy of sex reared its lovely head. The animals started to come on heat well before it would have been politic to point them at the bull and many of their desires and repressions would be taken out on any field visitors. They soon found out that satisfaction could not be found in each other or the dog, so anything else was fair game. Woe betide you if you bent over to cut the cord surrounding a bale of hay or straw as you would almost certainly end up with a randy animal mounting your back. Although they were far from full size, they were big enough to require that such sexual advances should be severely discouraged.

The desires of their flesh turned into a bloody nuisance when they discovered that mounting me did not give them much more of a kick than mounting each other. So they started to break out of their field in the search for something hairy and male. Joe, much to his annoyance, was not sufficiently attractive to the heifers and much as he ogled them over the hedge, they would ignore him, skip over the fence on the opposite side of the field and disappear towards the nearest bull that belonged to a neighbour.

This performance irritated me as I was rather proud of the fence. It was the only one, apart from my attempts to isolate the shit pit, that I built myself and it was a cunning three strands of barbed wire that weaved their way through the patch of scrub at the boundary of the field utilising trees and bushes as their posts, which meant that I could tighten the wires without fear of post sag. The vet was summoned after this first escape and all heifers that were on heat were expensively injected with his patent abortifacient to ensure that no calves were born while the mothers were still below the age of consent. The following day I laboriously ran sheep netting along my fence and felt smug. A few nights later, a couple of the heifers jumped over the top again. To give them their due, the attractions of the neighbouring bull were quite striking. Whereas Joe was cast in the Gary Cooper mould of solid worthy respectability, his equally purebred

Hereford rival was a John Travolta of a bull full of flashing eyes and dramatic poses, massive snowy dewlap and a rich and curly coat. The beasts always came home after the event as good as gold. A rattle of the concentrate bucket at the neighbour's gate and they would come rushing out and then it was only a case of running home fast enough to manage to keep ahead of them. If I was too slow, we would have an undignified scramble for the bucket in the middle of the road with cars piling up full of drivers sneering at the sight of such rank incompetence.

The vet was called again to administer his chemical knitting needle and I added an extra strand of barbed wire to the top of the fence and comparative peace reigned until the next time that they came on heat. I caught Number 1 burrowing under the fence Houdini-like, rather than going over the top. She took off across the twenty-acre silage field with myself in hot pursuit. She paused long enough to jeer at me puffing along behind her before leaping the hedge over to the neighbour's bull who fell upon her with a bellow of lust. It was obviously impolitic to come between a strange bull and the object of his desires, so I glumly kicked my heels until the amorous interlude was over before grabbing the heifer by the ear and leading her home to the vet.

I next carefully staked a strand of wire about a foot in from the main fence and this gave her pause for thought for another month before she escaped yet again. Things were obviously getting a bit tough as she left chunks of hair behind on the top strand of the fence. This time I raised the fence to about five-and-a-half feet and placed an electric fence in front of the barbed wire. Success at last. The fence would now have served admirably round a high security prison and the whole structure swayed and sang with every breath of wind. There was a sequel. For some reason she failed to abort after her last outing, or it may have been that she sneaked back home under her own steam after one of her nights on the tiles. Whatever the reason, she was found to be solidly in calf a few months later and produced a calf the following winter.

She was nowhere near big enough as a modestly sized eighteen-month-old animal for the rigours of calf production, but she produced a huge bull calf rather in the same way that you can open up one Russian doll and find another one inside just about as big. Her back end expanded like a boa constrictor dislocating its jaw to swallow a goat. After the calf came out, she gave it a disdainful sniff, snapped her back end straight back into shape and started to chew some hay. We let her feed the calf for a week or so and then dried her off and put her back with her fellow heifers. She never looked back and grew into a fine healthy cow.

The electric fence that proved to be the ultimate deterrent to her was the only form of fencing with which I could ever come to terms. Barbed wire and I could never get on together. One of the pleasures of living in rural bliss far from the pollutants of industry was that the air was clear and pure so long as you stayed upwind of the slurry pit. This meant that lichens grew on everything and that galvanised wire did not rust. There was a strand that was as good as new that was buried a foot deep in the trunk of an oak, where it must have been stapled half a century earlier. I never managed to use barbed wire without being damaged by it. Usually it would creep up on me and lovingly wrap itself round some delicate portion of my anatomy and hack out small pieces before it could reluctantly be persuaded to relinquish its grip.

Once, when chugging about in a field on the tractor, I came across a single strand of wire poking about a foot out of the ground. This was not a good thing as anything that juts out or is sharp invariably jumps out for a quick slash at the udder of any passing cow. I fixed this bit of wire to the back of the tractor as it seemed to be firmly buried and took off down the field. Wire came out of the ground and kept on coming. About sixty yards of it emerged and, once free, it decided to coil itself neatly up. The tractor was haring down the field being overtaken at every revolution of the wheels by a rapidly growing coil of barbed wire. It caught up with the tractor and hopped up on to the seat beside me and skilfully ripped my shirt to ribbons before jumping down again and wrapping itself tightly round the tractor axle.

I made it my vocation to remove as much of this fiendish wire from the farm as I could. The system of grazing practised was set-stocking which meant that the cows could go pretty much wherever they liked. Very few fences were needed to keep them in so I ripped down hundreds of yards of the stuff and piled it up behind the barn like a system of trench defences. By removing all the wire, I put a stop to the constant trickle of cows that would come into the parlour with damaged undercarriages. Nationally, the damage that barbed wire does to the stock it is supposed to protect must run into millions of pounds each year.

During one of the herd escapes, caused by some other idiot leaving a gate unfastened, the whole lot piled on to my neighbour's lawn after the morning milking. They all came back into captivity quite peacefully except for a bunch of real heavies who decided to demonstrate what a delightful spectacle a bovine Grand National would be by coming back over the barbed wire fence. All made it like Red Rum except for one of the cows who left half a tit behind. Joe thought that this looked quite fun and decided that he would have a bash at the fence. The sight of a short-legged bull with his testicles hanging beneath him like a couple of oranges in a string bag shaping up to a barbed wire fence is enough to turn the bowels of any watching male to water. Myself, my neighbour and my neighbour's dog all averted our eyes from the appalling spectacle. Joe came to a skidding halt in front of the fence and tried to look as if he had been joking all the time — an attitude which was rather spoilt by his ring getting tangled up in a barb and needing unhooking.

Electric fencing, on the other hand, works by bluff. Sure it can give you a bit of a kick but has never done permanent harm to anyone or anything, provided that you treat as apocryphal the tale of the elderly guest leaving the wedding reception for a quick pee and inadvertently choosing as his spot an electric fence which led to his demise through a heart attack. We had a very similar happening with the dog who was caught short in the middle of a bare field with an electric fence post being the closest approxima-tion to a lamp post that his town-bred eye could see. The complete outrageousness of the assault sent him howling back to the house with his tail between his legs and left me in weak giggles.

There is splendid and satisfying logic about an electric fence. One can go on extending it for ever provided that it does not short anywhere. All our cows respected it and, even after I had not bothered to replace dud batteries for a month, not one of them dared to touch it. The most effective and trouble-free fence was wired up to a mains fencer in the buildings, and since I was too mean to buy a sufficient number of battery machines, it had about a mile of wire emerging from it and meandering all over the farm. This system reached its peak during the Great Drought of '76 when all sorts of odd areas of scrub round the farm that would not normally have been grazed had to be opened up to provide grass of some kind for the stock, and electric fencing was the only way that order could be imposed.

We rigged a couple of magnificent spans across the farm lane from oak trees to telegraph poles and sent wires scurrying through hedgerows and under gateways. We also found half a mile of disused telephone wire and with this we shipped the current to the furthest regions of the farm. I was milking the cows one evening during the drought when they were behaving particularly badly. I was about to administer some drastic rebukes when I touched the handle of the cake dispenser and received a shock. One of the more impressive spans ran over the milking parlour and had sagged so that the wire was touching the tin roof. The whole building, the cows and, ultimately myself, were pulsating in sympathy. That particular system suffered a sudden collapse when the rains came and shorted it out in about a dozen places.

The drought was one of the natural disasters that we faced as dairy farmers but, head and shoulders above all others, was the Blizzard.

# Chapter Six

Let me describe a typical winter morning in the life of a scrub dairy farmer. It will be pitch black outside with the sleet beating against the bedroom window. The electric alarm clock that pulses and mutters to itself all night will suddenly erupt into its insidious buzz. I reach over and switch it off before snuggling deeper into bed.

A moment's peace and then a thump as my wife's foot makes contact with my posterior. It is 5.30 am in January and the bloody cows are waiting to be milked. I crawl out of bed and turn on the bath, gazing blearily out into the swirling darkness. Bathing in the morning before the day's cowshit has started to build up in layers on my skin is a pretty daft habit, but at that hour of the morning I certainly have not got sufficient initiative to change any of my habits. I collapse into the warm water fighting against the sleep that is as seductive as a glass of whisky to an alcoholic. What the hell did I ever want to keep cows for? Resolve suddenly hardens and I eject myself from the bath and into long johns, vests and sweaters and all the rest of the garments that would make Captain Scott look like a bathing beauty.

I totter down the stairs past snoring wife and children. The dog slinks guiltily off the sofa and tries to look as if he had spent all night on the floor and, much as he would like to help me with the cows, he is far too busy guarding the house. I make a cup of coffee in the hope that, for once, it will revive me, and peer at a copy of the previous day's paper. The moment can be delayed no longer. As always I have left the torch down at the buildings. I exit from the kitchen into night. Blast, it's freezing. The snow stings my skin. I turn left and feel my way along the wall towards the garage where my boots and overalls are. Why didn't I put the car away yesterday? I curse and rub my shins.

A slight change in the quality of the darkness and I am in the garage. I fumble for the light switch. The contrast between the blaze of the 40-watt bulb and the dark hurts my eyes. I empty from my boots a puddle of snow which has blown through the open door and pull on overalls that are stiff from the previous day's offerings by the cows. Out goes the light and I set off down the concrete lane to walk the 250 yards to the milking parlour. It is a question of feel: the feel of the concrete underfoot. Deviate from the circuitous route and you hit hedge or bog except at the worst corner where you would step down two feet into the stream. A slight hollowness underfoot means that I am crossing the bridge and that is the last main hazard over. There might at least be a star of two so that I could tell where lay the horizon. Careful not to fall over on the ice of that puddle. The stalks of some of last summer's nettles brush against my hand. That means that I am just below the dairy steps. I pull myself up and flood the interior with light. It is dank and cold, with the gleaming silver milk tank sitting in the middle with the paddle inside humming quietly as it stirs last night's milk.

I am now fully awake and fall into the routine of the morning. Taps off. They are left dripping to prevent freeze up. It is just as well that they have been left on as it has been freezing hard in spite of what the forecaster was saying the night before. Put a fresh filter on to the end of the milk line and put the line into the tank and check that the drainage tap is shut. Routine now but routine that has been bought the hard way. This particular bit of the routine is performed twice a day every day of life. It is all too easy to leave the tap open or the line out of the tank and return to the dairy after milking and find that I have poured the lot straight down the drain. It is seventy pounds an error but it took half a dozen such errors before the routine was safely embedded.

I go through the door into the parlour not noticing the graffiti and the murals painted on the wall by all and sundry in an attempt to brighten the place up. I go to the back of the parlour and yell out into the dark for the cows. Down into the pit to set up the milking equipment. Damn. I had forgotten to drain one of the clusters and it and the jar are now solid with ice.

I go back to the dairy for a bucket of warm water and spend ten minutes melting it out. On goes the radio. I never did care very much about the price that carrots are fetching in Leeds. Rustles and coughs from beyond the parlour door indicate that the cows are beginning to gather in the collecting yard. An impatient thump on the door heralds the arrival of Number 4. She has already smashed down the door once when, in her opinion, it was not opened fast enough.

I go out the back of the parlour into the yard to chase up any stragglers. A quick look into the bull pen — the calves are still alive. I am not quite sure whether to be glad or sorry. Another quick look into the cubicle shed to ensure that all the cows have gone down. Number 50 is stuck again. She appears to be having some sort of a problem with her back legs at the moment and has difficulty in hoisting herself to her feet. I may have to call the vet in to her but I doubt if he could do very much for her. I go in and haul out a six-foot beam which I use as a lever for her particular problem. I slip it underneath her belly and heave. It is just the added bit of impetus that she needs and she shoots to her feet and steps delicately out of her cubicle and down towards the collecting yard.

It has not thawed for two or three days now and the whole yard is covered by a slick of frozen dung and urine. The buildings are built on a slope, and so the yard is highly treacherous for both man and cow. The last bit of hill down to the yard is the steepest and 50 solemnly plants each of her hooves on the ground and skates ponderously down over the ice. She begins to rotate half way down but, experienced animal that she is, corrects her direction and ends up with a gentle thump against the gate.

As usual, I watch this performance and curse myself for forgetting to spread salt on this slope. The cows seem to quite enjoy it and even form queues during the day so that they can slide down, but it is surely just a matter of time before one slips and hurts herself.

I check the silage face. Everything seems to be in order. The clamp that they are munching their way through at the moment is about eight feet high, and so I had better fork some down from the top to prevent an avalanche that might land on top of a cow and suffocate her. Christ it's cold. I go back down towards the parlour. Joe is standing outside the collecting yard so there cannot be any cows on heat. I give him a quick scratch along his back as I go past and shut the gate to keep the cows in the yard. It is full of cows quietly cudding with steam rising from them and swirling round the light bulb. There is a quick flurry by the door to the milking parlour where 4 remonstrates with another cow that she considers to be intruding in her space.

I go round and up the parlour steps once more and slip the milking apron over my head before pressing the buttons that start up the machine. As always it bursts into immediate life. It cannot last forever. Some icy Sunday morning, the machine will fail to start and then I will be in big trouble. Through the dairy door and into the parlour. I fling open the door to the collecting yard and dive straight into the pit to avoid being trampled to death in the rush of cows.

I give them some concentrates. What on earth is 46 doing in the first batch? She is not a pretty sight at this stage of the morning. 46 has a little problem. She has an irrational hatred of the cubicles. If I go down to the buildings on the most miserable night of the year, the rest of the stock will be tucked up cosily in the straw of the cubicles but 46 will be sitting out in the most exposed and windswept position, probably in a pile of slurry. I have tried everthing I can think of to sort her out, except a psychiatrist. She has been tied up in the cubicles, fed in the cubicles and even been shut up overnight in the cubicles; various fiendishly subtle ploys have been used to persuade her to sleep in the beds rather than in the dung passage, such as strewing the passage with rocks or flooding it. She had no objection to a waterbed and treated the rocks like foam rubber.

So washing the udder of 46 is a major undertaking. It is a question of deciding how far one should go. There is a clear Plimsoll line round her body above which she is clean and below which is a thick encrusted coating of straw and dung that rustles and clatters when she moves. The only way to clear this would be by going over her with an electric razor, which would take an hour or more and probably lead to pneumonia. So I would remove as little of this muck as I had to, to ensure that milk and not slurry ended up in the jar. If she had been up at the silage face for an hour before milking, the job of cleaning her udder would be easy as the slurry would have dried and could be removed intact and lie on the floor looking like a sinister ghost udder or an old leather glove.

It is a pleasantly routine milking. Each cow gives out a good kilowatt of heat and the parlour soon warms up and prevents any further chance of freeze-ups. There is the odd flurry of excitement as the odd cow tries to pre-empt another's place in the queue, and the usual hassle with the bantams. After milking, I wash down the parlour and then scrape out the cubicles. The cows are let out into the lane to try to prevent them obstructing the tractor. A pretty vain hope. The cow's day is rather dull. Eat, sleep and stand around, and the chance to dice with death by skipping round the tractor is hard to resist. It is a fairly fatuous occupation trying to scrape when the frost has welded the slurry to the concrete. Sometimes the odd section will become detached like an ice flow but will pile up again after a few yards and the scraper will skid over the top. The tractor itself has difficulty in climbing the hills of the yard, particularly when asked to push slurry before it. The pit, of course, is full and a mountain of rock-hard frozen dung has built up at the entrance. If I take a run at it, the pile might lurch out into the pit a couple of inches, but the scraper is quite likely to bend.

The normal pattern of a winter day would continue. Sometimes normality flew out of the window. Never more so than in February '78. That February was not the first time that we had had hard weather. It was always a bit inconvenient but not really unexpected during the winter. The main drawback of frost was that it made the farm lane impassable. We were very lucky in that a few years before we took over the tenancy of the farm there had been installed half a mile of concrete down its length so that the tanker could be enticed in to empty the bulk tank. The regular driver knew it well, but relief drivers would arrive, blaspheming, at the bottom of the hill leaving the hedge bank studded with their wing mirrors and mudguards which were sloughed off on their way down.

The builders of the lane had the bright idea of sloping it towards the middle so that surface water would run down the centre and not scour away the banks at the edge. Surface water trickled down almost all the year round as the lane was five feet below the level of the adjacent fields. The idea behind this slope was shown to have flaws every time that it froze. The water in the centre would turn to ice and it would build up in layers as it continued to trickle out of the fields. In a surprisingly short space of time, the entire length of the lane would be a sheet of ice, blocking it to all but those who felt that they needed practice for the Cresta run.

Snow, if anything, made it better rather than worse because there was at least something there for tractor wheels to grip on.

This was needed because if the tanker could not come to the milk, the milk had to go out to the tanker, otherwise it would have to be poured down the drain. When snow came, all the local farmers dug out their emergency tanks. They came in all shapes and sizes designed to be put on the back of a tractor or trailer and hauled up to the central collection point where the gallant tanker driver would be waiting having battled his way through half-inch snow drifts in order to pick up the milk.

My tank was made of exceedingly fragile glass fibre that needed constant repairs in order to keep ahead of the innumerable cracks and holes that leaked the milk out on to the snow. Its use was further complicated in that the weight of the milk it would hold completely overbalanced the tractor and meant that it had to travel backwards. If it went forwards, the front wheels rose in the air like the legs of a praying mantis and made progress rather difficult. Piles of fertiliser bags on the bonnet effected a temporary cure until the heat from the exhaust melted the plastic or burned through the paper and created an expensive trail of granules between the farm and the collection point, and the wheels rose gradually skywards.

The wise man might say that it would have been more sensible to take two trips, but apart from the extra work involved through doubling the length of the journey, there was always a queue of farmers waiting to unload their milk, and even had the tanker been willing to wait while I made the second trip, it was doubtful if he would have had room to take the milk. I also found certain difficulties in transferring the milk from the bulk tank in the dairy out to the emergency tank on the tractor. It was one of those times when a bit of lateral thinking would have saved me a great deal of trouble. Being brought up in the logical school of thought, I messed around with syphons and backing the milk up through the pump until a neighbour came down to observe my progress. He seemed a little surprised at my Heath Robinson method. 'Why,' he asked, 'don't you just pull out the bung in the bottom of the tank? The tank in the tractor is lower than the bung hole so it will pour straight in.' I looked and wondered why it had never occurred to me. The milk ran straight down a sheet of corrugated iron into the tank and took ninety second to transfer rather than half an hour by syphon.

This gave me more time to sit and get cold on the back of the tractor at the collection point. It was a bit like a garden party really. Farmers who hardly saw each other from one year's snow to the next would meet, discuss their snow problems and the general state of the nation and moan because they had to hang around so long for the driver when they should have been doing more important things like keeping warm back home.

Anyway, one Saturday we were all waiting in about three inches of snow at the collection point having a good gossip when it started again to snow and soon turned into a full-scale raging blizzard. It was close to midday and the trip back to the farm was made into the teeth of the storm, which on a cableless tractor was no fun at all. That was the last time that we were to see the tanker for a week. That evening I struggled down to the parlour for milking with the snow still falling hard and the cows looking decidedly sorry for themselves. I thoroughly strawed them down in the cubicles after milking and, in an unprecedented gesture, gave 46 a bale all to herself in the corner of the yard. I took the precaution of shifting the tractor up to the lee of the house to give it some sort of protection against the snow and to point it down the hill, as I could see that there might be some problems in getting it started in the morning due to blown snow forcing its way into all the crannies of the engine. We went to bed with the blizzard still raging outside.

In the morning a few hours later, the wind was still buffeting the side of the house. I made my morning coffee with more than the usual degree of dread, donned the many layers of waterproof and protective clothing with care and went to open the door to venture forth. Our doors, unlike any others that I have met, opened outwards rather than in and on this morning were immovable. I scraped a hole in the frost of the window and looked out. Nothing. I scraped a little harder until I realised that what I was looking at was not frost but piled-up snow. I climbed on to a chair and pulled down the top of the window and launched myself out over the drift and slid down to its bottom.

Although it would normally have been dark, I was a little later than usual — it had been a rather unusual morning so far. There was something decidedly unnatural about the landscape. It was still snowing and everything was in various shades of grey as if all colour had been drained away. The farm looked as if a giant comb had been raked over it, scouring a few patches clean of snow and piling it up behind any obstruction in its path and all shining with a strange, translucent grey light. Down in the valley, along the stream, the trees were standing black, the wind stripping the swirling snow from their branches as it touched. And it was cold, bitterly cold, the blizzard driving into my face.

I pulled myself to my feet and started down the hill to the parlour. The first thing that I noticed was that the tractor had disappeared. Where it had been carefully parked beside the house the night before, there was now only a mound of snow about eight feet high. Then I realised that I might have problems in getting down to the parlour. The lane was blocked by huge drifts of snow. At the top of the lane, there was a gateway. Across it had built a drift of quite awesome beauty, had I time to sit down to admire it. It was like an enormous wave that had frozen at the moment of breaking, with a graceful sweeping curl coming to a point that was continually building up and being swept away again by the wind like the vapour trail of an aeroplane.

It was obviously not the moment to rhapsodise and I snuggled deep into my jacket and set off down the lane. It took a good fifteen minutes to travel those two hundred yards. There were only about two or three drifts that had built their way completely across the lane. It was still possible to walk round them by the hedge, but none of my training had taught me how to get through a drift of powder snow that was as high as I was. I spend several minutes of useless floundering until I discovered the technique of forcing my way into each drift as far as I could go and then somersaulting over the crest and slithering down the other side.

I reached the parlour and dairy and found that things were not much better. Full frost precautions had been taken the night before, which was just as well since the air inside the parlour was saturated with a fine mist of blowing snow that hissed and rustled as it fell. I climbed up the ladder into the cake loft to shovel concentrates towards the dispensers and found the floor a foot deep in snow that had blown in under the eaves. Lining the beams there were about fifty or sixty birds — thrushes, blackbirds, tits, finches, robins, pigeons and even a snipe — sheltering from the weather. The rafters were only about four feet off the floor and I had to duck under them to get to the cake but the birds just sat and watched me out of their beady eyes, immobile and completely silent.

I came down to bring in the cows. There was a drift about five feet high blocking the way into the collecting yard. It was the first time that I put my hand to the shovel and I cleared a passage through for the animals. Fortunately most of the yard was being scoured by the wind and, after an intensive five minutes' work, I managed to persuade the cows to leave their nice warm cocoons of straw in the cubicle shed and sprint the twenty yards down to the collecting yard.

Milking completed, I moved the electric fence that ran across the silage face closer so that the cows could eat their fill without wasting any of it. The top of the pit was covered by snow but the face, warmed by fermentation, was a vivid splash of green that stood out from everything else. There, too, every perchable niche where a cow had grabbed a mouthful of grass was filled by a huddled bundle of feathers.

The rest of the day was a bit of a wipe-out. All the stock were indoors and in no immediate danger — unlike the miserable sheep. The conditions outside made work almost impossible and most of the day was passed in trying to prevent any fresh drifts forming in awkward places. Some time during the day the media told us that the snow had stopped falling, but there was no way we could tell as the wind chased the drifts across the farm and filled the air with snow.

Ominously, towards evening, the foot-thick wooden beams that held up the corrugated iron roof above the silage barn began to crack and bend under the weight of the snow that was piling up. We propped them with lengths of timber and surrounded the props with electric fence wire in order to keep the cows from knocking them down. We returned to the cave of a house for the night, the windows piled high and dark with snow.

The following morning the wind had died away and it had stopped snowing. It was time to take stock of the situation. The first priority after milking was to try to clear a path so that the milk could be taken out. Although it was a little optimistic to expect the milk tanker for a day or two, we would be in deep trouble if we were still snowed in when it arrived at the collection point as we had facilities on the farm to store no more than a couple of days' milk.

The first thing to do was to try to locate the tractor and start it. Finding it with a probe took only a minute or two, digging the engine clear and starting it took a little longer, but start it eventually did and shot a mighty gout of steam from its exhaust pipe which froze instantly on coming in contact with the air and drifted slowly down the lane, a sparkling, dancing cloud shimmering in the sun.

I drove the tractor out of its blanket of snow and sent it with a soft 'whump' into the first snow drift on the way down the lane. The snow cascaded down over the bonnet and it stalled. Heaven preserve all wise virgins who thought it smart to park at the top of the hill to allow a good run down. Two hundred yards of snow drift was not an enticing thought to tackle with a shovel and we cast about for alternatives.

Charles Gore

A streak of genius was the utilisation of our fifty-five cows as the sharp end of our snow-clearing operation. Grumbling with protest, they were winkled out of their cosy cubicle shed and pointed at the snow drifts. They balked. I yelled and jumped up and down and, reluctantly, they turned to their duty. Number 4, as always, was the star of the show. Most of them worked fairly well in about three feet of snow but lost interest very quickly once their bellies had grounded and they lost traction. Number 4 plunged, snorting with excitement, into the high spots of the drifts and I forced the rest of the reluctant herd through the gap that she created. We ran the cows up and down the lane a few times and within minutes there was a neat alley of hard, compact snow weaving its way through the drifts just wide enough to take the tractor which chugged its decorous way down to the buildings.

The next step was to explore the possibilities of taking the tractor up the lane to the road. The first exploration gave us some idea of the extent of the problem. The lane was blocked to the level of the hedge, ten feet high, all the way up. Patches of the fields had been blown clear but in the lee of water troughs or hillocks, massive drifts had formed. We walked up to the road along the top of a hedge and again were confronted with these enormous drifts. Where the lane ran east–west, it had been scoured clear. Where it ran north–south, the wind had dropped the snow over the edge of the hedge and there it had quickly built up into drifts. From one of the larger of these we could look down to the top of the telegraph pole that was poking through. We worked out a rough route from the buildings, winding its way up the fields round the drifts, up to the road, and we started to dig.

The first dig was about an hour to break out of the yard as all the gateways were blocked. Once through these drifts we were in comparatively open fields. Comparatively, because the snow was only two to two and a half feet deep. We discovered that the tractor would only go so far in this depth before it stopped. Initially we dug down along the path that its wheels were expected to take in the hope that it would gain added traction, but this seemed to make little difference. The problem, we found out, was that hard snow built up under the sump and lifted the whole machine off the ground. This cut down digging by half, as all that was needed was a narrow trench along the line of the sump and the machine would go beautifully.

We pushed our way out to what we hoped was a clear spot on the road. A couple of hundred yards further down, we could hear our neighbours at work on the same mission. They might as well have been on the moon as far as making contact was concerned. Once on the road, we were faced by a series of massive drifts before which our resolution failed and we turned back into the fields and burrowed our way across country for another quarter of a mile before breaking through to the road once again. This time we were well away and thrust our way along the track that had already been carved out by others and made it to the collection point. It was a bit of an anticlimax as the estimated reopening of the main road was four days away.

We retraced our route to the farm. Once inside the farm boundary, our hard-won track meandered its way for nearly a mile before reaching the buildings. Quite an achievement on only seventy acres. Just inside the boundary was a temptingly smooth stretch of virgin snow that ran straight down to the buildings. We walked down over it to test for depth and 'tractorability'. It seemed to be fine and away we careered down the hill, smashing our way through the snow. The awful truth dawned rather more than half way down. The tractor was rather heavier than we were and was sinking rather more deeply into the snow. It ground to a halt with the snow lapping at the base of the driver's seat. Once again a long wearisome dig, to extricate the machine and retrace its path to the top and then round the track to the buildings.

Back there, things were standing up quite well. With the absence of wind, the drifting had stopped and the cows had beaten paths throughout the yards. The props were holding the roof up without much difficulty, and the only problems were wondering what to do about the slurry that was lapping six inches deep in the dung passage of the cubicle shed — its normal outlet blocked solid by snow — and ensuring a constant supply of water as the pipes had all frozen solid. We milked the cows, beginning to get the first case or two of mastitis probably brought on by the insanitary conditions and weather stress, and dragooned every container we could find into service to store the milk until we could sell it again. The forecast was for thaw and we returned to the house feeling that we had probably lived through the worst of it.

Disaster was apparent in the morning. It had indeed started to thaw in a big way, but water from melting snow trickling down the roof supports had shorted out the electric fence. It appeared that a cow must have gone to have a scratch on the prop and brought the pole down. This resulted in a section of the roof, thirty yards square, crashing down in front of the silage face leaving a twisted mass of corrugated iron and shattered timbers.

Through the devastation strolled the cows, chewing placidly at their cuds and appearing to be totally unmoved by the change in their habitat. I brought them in to be milked as water was beginning to drip through the roof of the cake loft where I had not got around to clearing out the snow. The first batch of cows came in when, with a sound like an erupting geyser, a great wave of slurry came gouting through the parlour door. I vacated the pit with alacrity. The dam blocking the cubicle shed had burst and a torrent of shit bounded down the cow-made path and filled the pit in the parlour to a depth of two feet. Baling out freezing slurry at 6 am must come pretty low on the list of ways to pass an hour.

I cleared the pit and continued to milk the cows. A couple of them had very nasty cuts obviously caused by the collapse of the roof, but we seemed to have been lucky. Or thought we were. We were one cow short. A rapid check through the herd showed that simple 6 was missing. We looked in all the usual places. Cows had been known to go missing before and turn up again — for example, shut in the feeding passage of the bull pen after half the countryside had been scoured for signs of their presence. Number 6 was not in any of the usual hidey-holes and the only place left unsearched was the daunting pile of wreckage left by the falling roof.

We picked up the well-worn shovels and cleared off the snow. After an hour or so, we were down to tin and still no sign of 6. A couple of bulges in the tin were prospected. It was a surprisingly difficult job as the builder had tacked the tin to the timber at a rate of a nail per inch. Number 6 was under the third bulge. She must have been quietly cudding away when the roof fell in — she might have been the one enjoying the scratch that had knocked the prop down, which would have been some form of poetic justice. At any rate a beam had caught her across her rib cage and crushed her chest; she must have been killed instantly, before being buried under tons of snow and rubble.

The rest of the day was spent in the preliminary clear-up of that mess, removing the remains of 6 to a less public corner of the yard and repairing all the burst pipes that became apparent with the advent of the thaw. Our track through the fields had turned into a mini M1 that wound its way back past our farm through hedgerows and was part of the supply route of several farms to the south. So we had a trickle of neighbours bouncing through, prodding the remains of the silage barn and comparing disaster stories. Some of the tales made our little problems seem like a picnic. Flocks of sheep vanishing without trace. Others buried up to their necks in snow and having their eyes pecked out by crows. A complete herd of cows being buried under a collapsing cubicle shed and one farmer who took two days to dig a way through to a shed where his cows were shut in waiting to be milked.

The helicopters started to buzz overhead on their missions of mercy, like dropping loo paper to those who had run short. We were given a rather predatory inspection by a giant American Sikorsky looking like a flying block of flats that had been sent over from East Anglia. It hovered for five minutes above the collapsed building before it decided that a herd of cows diving into the shit pit and going generally berserk with terror might be partly its responsibility and it thundered off.

The milk situation was becoming fairly critical. Fortunately my emergency tank had suffered one of its periodic collapses and decanted 150 gallons of milk into the snow but, once repaired, it gave another day's storage space. Everything else from buckets to water troughs was now brimming with the liquid. It is a great shame that you cannot turn cows off during natural disasters and strikes. The following morning, the plug was actually pulled on the bulk tank and we all stood in a solemn circle and watched 300 gallons gush down the drain. In a way it was quite a relief as it meant that the worry of storage space was one worry less. The day after that, with the thaw still continuing full blast, there were actually rumours of an approaching milk tanker and half the farmers in the county loaded their tractors with milk and chased this phantom through the now sodden snow drifts. The dairies had been caught with their pants down just like the rest of us and, as one would grind into action, another would break down, marooning its fleet of tankers out in the wilds without a home to come back to.

The track across the field was now showing deep, muddy ruts and so we started a determined attempt to clear the lane. I stuck the slurry scraper onto the back of the tractor and charged up the lane as far as the tractor would carry me before dropping the blade and reversing back down, sweeping a great wall of water and snow through the yard and down to the stream. A neighbour coupled his tractor to mine and, together, we hauled the remains of 6 up the lane to the crossroad where the kennel van could pick her up. She gave us a final service by clearing a path through the snow that allowed a single vehicle to get up and down. The massive drifts that blocked the road were tackled next. It had been thawing fast for five or six days, but there were still such massive piles of snow around that it took six weeks or more for all trace of them to melt away from the hedgerows.

We dug tunnels through the drifts just about the width of a tractor and life slowly began to return to normal as the snow melted. Such vast quantities melting at such a speed caused quite severe flooding. Our own little stream was a swollen torrent, spilling over the bridge and even carrying away some railway sleepers that were stacked beside a nearby gateway. The ice flows that sailed down it piled up forming dams that further backed up the water.

The stream overflowing the bridge was nothing new. It tended to do so every time that there was any serious rain. Its overflows came in two varieties. The normal entailed an uncomfortable wade across the bridge, while the much rarer boot-drowner came about when the height of the water going across the bridge was above the level of my gumboots. The thaw was the worst boot-drowner we suffered because of the frigid temperature of the water that did the drowning.

A curious legacy of the snow was revealed when the drift that tried to block the entrance to the collecting yard was eventually cleared. At the bottom were a couple of polythene buckets, good stout ones, that were used for feeding calves. They were both squashed completely flat.

1978 was not the only year that we had snow, although in other years it was usually decorously distributed throughout the winter rather than dropped on our heads in one dirty great lump. 1979 was again a bad winter, but down our way we had all learned our lesson from the year before and were ready for it. When the snow started to fall, we took to the lanes with our tractors and scrapers and knocked down the drifts before they had a chance to start to form. We were damned if we would be bottled up again.

During that second hard winter we had a bit of a problem with the horrible heifers that had spent the previous summer in their orgies with the bull next door. We had a ten-acre field that had a shed in it, and since we were determined to keep the blighters as far away from the bull as possible, that was where they spent the winter, far from the bright lights of the buildings. They received a few concentrates but, for forage, they had to rely on pickings round the field that the cows had left behind; hopefully, the pitter-patter of their tiny feet would make rather less of a mess than those of their elders and betters.

With the advent of the snow, the situation changed somewhat. Their larder was now covered in a white blanket. Being exceedingly thick and not reindeer, it was too much to expect them to scrape for lichens in the approved fashion and so I had to supply them with a couple of bales of hay and straw a day. Naturally, access to their field was blocked by drifts. The only way that they could be supplied was by running the tractor up to the hedge, precariously crawling along its bonnet with a bale in each hand and then swaying carefully along a couple of planks placed from the hedge-top over the drifts to their shed.

This worked quite well for a while until the animals grew to expect me and decided to anticipate their grub by coming to get it rather than waiting for it to come to them. In their excitement they would all pile up the plank towards the hedge and then slip off it into the deep snow and, looking pathetic, wait for me either to dig them out or pull them out. I heard of one man whose horse was caught in such a drift with only its head poking out. It stayed there for a week while he fed and watered it by crawling over the snow on a ladder. Then the snow melted and the horse stepped out on its way, none the worse.

Conditions like these made for adaptability. We made snow-shoes out of wire netting to help us get about. I dug out a pair of skis that had been unused for fifteen years; these worked admirably until coming downhill at some speed for the first time I realised that I had forgotten how to stop. I was faced with one of those awful dilemmas with which one's life is strewn. Should I chance the stability of the drift that was covering the stream and sweep gracefully over it? Or sit ignominiously down on my bottom before I reached it? I gambled on the glamorous option and finished chest deep in snow with the water gurging up to my knees and the skis still firmly attached. Freeing myself took a fairly frantic ten minutes before I floundered my way back to warmth and safety.

The bantams were not up to dealing with the cold weather. I recalled their presence during one blizzard or another and went to see if they were all right. They perched on a hay rack in a shed at night and the snow had built up to their knees so that they looked on the point of death. I nearly decided to let nature take her course and pretend to my conscience that I had not reached them in time, but I relented and hauled them all into the milking parlour to live there for the duration. They thoroughly approved of this and hopped their way up the pipework to roost just below the roof. Between forays down to rob the cows, they would amuse themselves by dropping their odiferous bombs on my head as I worked underneath. They were ejected again as soon as the snow began to melt.

Other birds depended on us for hand-outs during hard weather to enable them to survive. There would normally be little other than tits and sparrows scrabbling by the kitchen for scraps, but when the snow came we would get partridge, snipe, woodpecker, pheasant and a dozen or so other species that would clamour outside the window for food.

The deer used to come visiting when things became tough. The Devon red deer has a far softer life than his cousin in the North because he lives on far richer land. They were by far the most spectacular members of the local fauna and, because of hunting and poaching, are not all that kindly disposed towards man. During the summer, we would regard them with mixed feelings. Granted it was very nice to have them around but it surely was not necessary for a couple of dozen of them to graze all over our silage field the day before I was going to cut it. During the blizzard when the snow lay round about deep and crisp and even, a group of hinds came down to the back of the milking parlour every morning and waited for me to scatter pounds-worth of cattle nuts over the snow for them. They would delicately nibble at them watched by myself wearing a soupy smile, with the cows leaning over the gate bellowing in their disapproval at this waste of good food.

Apart from such abnormal times, the deer tended to be very wary except, for some reason, when I was on a tractor. Then they would amble by without showing any interest. The glorious exception took place one November when a bit of late digging was being done in the vegetable patch by the house. The digger was working happily away when he heard a strange snort just behind him and there was a red deer calf sniffling at him just over the fence.

All and sundry were summoned and we all went 'Ooh, how sweet' under the animal's benevolent gaze. The expert neighbour then ventured the opinion that the calf was really too young to be expected to survive the coming winter and, moreover, its mother had probably come to a sticky end, otherwise it would not have been wandering around by itself. It was therefore our duty to capture it and feed it over the next six months. Catching a thing as small and as well disposed to us as this calf did not appear to present too much of a problem. We formed ourselves in a semi-circle against the hedge with the calf in the middle and moved in. The calf waited until there was about eight feet between us before it gently skipped through the gap, stopped a couple of yards away and looked benign. After a couple of similar attempts with similar results, we decided to trap it in the yard and plodded down the lane with the calf following us. We manoeuvred it between us and the desired pen and tried to shoo it gently forward again. It skipped between us and when one member of the party made a wild grab at it and tripped and fell flat on his face, the deer stopped and sniffed at his face before moving off the customary two yards.

It became bored by us after an hour or so and sloped calmly away into the woods. Apparently the complete undrivability of the West Country red deer came about through evolution. It makes a nice story anyway. The deer has been hunted with hounds for centuries by the local farmers and landowners who tend to conserve them for that purpose. During the nineteenth century, there was a period of a generation when the hunts lapsed and the deer were shot unmercifully as they were considered to be nothing more than vermin once the pleasures of the chase had been removed. People organised drives rather as the pheasant and grouse is driven today towards the waiting guns.

When hunting was re-established, very few of the deer were left and those that had survived were the ones that had broken back through the line of beaters instead of running away from them towards the guns. Their descendants now thickly populate the district and have retained this trait when faced with man on foot rather than on horseback. That little calf refused to be driven anyway.

The deer's eating habits became even less popular during that other great climatic upheaval — the drought in 1976. Agriculturally it started off as a very good year with a nice early spring. An early spring is a beautiful thing because as soon as the grass starts to grow you can get the cows off the unpalatable rag-ends of the previous year's conserved feed and out to the fields. It was the only April that we had when the hooves of the cattle threw up dust rather than mud. There was enough moisture in the ground to provide a splendid crop of grass for the first silage cut, although it was the worst-quality stuff that we ever made because as soon as we went out to cut it, the heavens opened and it poured with rain until we had brought all the grass in. Then the rain stopped and did not start again for four months.

The system of grazing that we practised had the effect of masking the effects of the drought from someone as inexperienced as myself. We set-stocked, which meant that the cows wandered where they liked, and this made it difficult to determine when they were running low on grass. If you switch your cows from paddock to paddock or keep an electric fence moving just in front of them, you know very well when the grass fails to regrow at the right speed. Our cows sat contentedly in the sunshine throughout the summer. There were a few signs that things were not quite as they ought to have been. The cows began to stroll along the tops of hedges chewing the leaves and they munched away at the bushes along by the stream where the badgers lived. The dung pats normally had a whiskery ring of uneaten grass round them as the cows, quite understandably, did not like the taste, but these whiskers were shaved flat. It was my first full grazing season and I had yet to find on what was normal and what was not, so I did not recognise these signs.

The first effect of the drought was that the water courses and ditches started to dry up and the frogs began to suffer. Between the house and the buildings was a ditch in which the love-lorn amphibians congregated each spring. I first became aware of them one dark evening when on my way to attend an unexpected confinement. Everybody knows that frogs croak but as a relative newcomer to country living, I had only heard them at it when David Attenborough had been around to poke a microphone under their chins. Twenty of them in full cry raised the hairs on the back of my neck until my relieved torch-beam identified them as the source of the racket.

When the frogs first started to fornicate and croak, it was an infallible sign that we were due for more snow. This would chill their ardour until it melted and then they would be at it again with their eyes bulging above the water line twisting themselves into great writhing piles while gently filling the ditch with spawn. There would always be a solemn heron in attendance at the nuptials eager to snatch any unwary lover who allowed passion to overcome caution. This heron would be up and down all day. The main fornicatorium lay beside the farm lane, and the heron would lumber into the air like a jumbo jet every time anyone went up or down it.

By about the middle of April, the jaded frogs separated and crawled off into the undergrowth to recover and leave the spawn to mature. During the drought year, their ditch dried up and I found myself being forced to carry bucket after bucket of spawn down to the stream in order to ensure a rising generation of frogs.

The heron was about the only beast that seemed to thrive that summer. I surprised him a few months later in one of the pools in the stream which by this time had virtually ceased to flow. The bird had such a gutfull on him that he ran over the field like an ostrich for fifty yards before he attained sufficient speed for a take-off. I went to look at what he had been guzzling and, to my astonishment, the half-empty pool contained several fairly substantial trout which only went to prove the high nutritional value of a leaky slurry pit.

I spent a happy afternoon wandering up the stream with a bucket and secured about twenty-five trout up to about ten inches long plus assorted loaches, barbels and millers thumbs and one surprised crayfish. To start a fish and chip shop was a bit immoral under these circumstances, so the fish were parcelled out around the water troughs on the farm and they seemed to thrive at a stocking rate of about four to a trough. There were some initial setbacks. For the first few days both the cows and the fish took a fairly dim view of cohabiting. A dreaming cow would amble up to a trough for a peaceful drink just as a trout would rise to a fly and

107

they would have a mutual fit of palpitating terror. A few of the dimmer fish were a bit carried away with enthusiasm in their pursuit of quarry and were apt to leap from the trough to the delight of the rooks, if not to themselves. Otherwise they all survived perfectly happily until they were returned to the stream the following autumn.

It was almost a year later when I was repairing a leak in the huge concrete trough that provided the cattle with their only source of water during the winter, that I came across a fish about a foot long which I had forgotten. For all I know he is still there. He was doing such an excellent job in keeping down all the creepy crawlies that tried to colonise that trough that I left him to it.

For about a week or so during the height of the drought, every time I went into the yard there was a curious high peeping sound that was just on the edge of the hearing. It was the sort of noise that could have been just my imagination or the Eustachian tubes readjusting themselves. I put it down to the consequences of the abnormally high temperatures. I was in the parlour ministering to one of the cows one evening when she sunk her hoof briskly into my guts. This was, of course, always an occupational hazard, but I had come to expect some sort of a warning which in this case had not been forthcoming. After picking myself out of the various little nasties that littered the floor of the pit, I discovered the source of the peep. Sitting on the gate at the end of the parlour, peeping away about six inches from the end of the cow's nose, was what was very obviously a young cuckoo.

I tiptoed gently towards it so that I could wring its neck and set the cow's mind at rest. Just as I was stretching my hand out to grab the bird, it exploded into flight like a demented frisbee, causing clusters all over the parlour to clatter to the ground like autumn leaves before it crash-landed a couple of yards away on the edge of a water trough.

It continued to hang around the yard for about three weeks keeping up this ridiculous noise and having a continual shuttle of hedge sparrows stuffing insects down its gaping throat. The only time that the shuttle stopped was when I or a cow came too close to whatever corner of the yard it was skulking in at the time. Then it would freeze until we came within a couple of feet of it when it would erupt into its manic flight, causing near heart failure to its visitor. It disappeared eventually, still peeping, and I energetically scared away all cuckoos that I saw the following year to try to prevent a repetition.

The long summer of the drought wore on. We were insulated to some extent by the sort of land that we farmed. The heavy soil retained its moisture for far longer than most types of land and kept the grass growing. We were also helped by having both main and spring water. When the springs dried up, we were able to rely on the increasingly spasmodic supplies from the reservoir. The state of the cows' lactations appeared to have been an advantage at the time. Joe had not yet pulled the calving dates very far back from the autumn and so most of the herd was giving little milk, if not actually dry. Eventually even our fields turned brown and did nothing more than sit and cook under the broiling sun.

At that point we erected our maze of electric fencing and began to strip graze what should have been our next crop of silage. This had grown quite normally for a couple of weeks before the grass had realised that something funny was going on and had halted leaf growth and concentrated instead on throwing up a long and virtually indigestible stalk crowned by a seedhead. Cows that began to calve were shunted over to this field and were allowed to chew unenthusiastically at this stuff; at least it was food.

The drought finally broke in September with a thunderstorm to end all thunderstorms. The ground had been baked dry for months so the water ran straight off the top, turning the lane into a raging torrent. It tore great slabs of muck from the yard where it had lain and cooked for weeks, before carrying it down to the stream which scoured half an inch of concrete off the surface of the bridge.

The financial results of the drought were unexpectedly severe. I had been able to carry the losses of the blizzard without too much difficulty but, during the drought, I had lacked the experience to realise to what extent the cows were being underfed and this showed up in the subsequent lactation when, across the board, they all gave considerably less milk than they should have done. Another unexpected effect was that it turned the drainage of the farm upside down. Where springs had appeared for generations, they went underground and re-emerged scores of yards away. An enormous pond had been dug in front of the house for ornamental, irrigation and anti-fire purposes and this lost four of the five springs that fed it and dried up to a couple of brackish carp-filled puddles. When the rains came, these lost springs never came back and presumably burrowed down rather than up to quench the eternal fires in the centre of the earth.

# Chapter Seven

After a couple of years as a dairy farmer, I made the rather embarrassing discovery that the less management of the cows I did, the more money they yielded for me. Management on a stock farm means making the animals do what does not come naturally. Otherwise they would not need to be managed at all. By running the bull with the herd at all times, the cows had already gone a long way down the road towards self-management and the set-stocking grazing system took them even further. The few times that I stuck my oar in such as, for example, keeping Joe out of the herd for a few weeks to try and push the calvings back towards autumn, always seemed to backfire on me in some way. That time Joe sulked so much that when I wanted him to start serving again he said nuts, and there followed a period of great uncertainty before he got into the swing of things once more.

So I moved in the opposite direction and turned the management of the farm over to the cows themselves and tried to foster in them a spirit of independence and responsibility. After all, I thought, the cows knew a damn sight more about what it was like to be a cow and what they wanted than I did. If I provided enough food for them and let them get on with their lives, they should become a thoroughly contented and, therefore, profitable herd.

The system that evolved suited me very nicely as my own function declined to becoming little more than acting as handmaiden to the cows. I took milk from them when their udders became uncomfortably tight and acted as lavatory attendant. It did not suit all my advisers and financiers. Businessmen are supposed to control all aspects of their operation. The cows ran my business and were not always good at passing on to me their

111

motives for running it in the way they did. Granted they produced profits for me, but all experts and bankers want to know how and why and they could hardly be expected to lend money if the responsibility for its investment rested with a cow.

I had originally banked with a delightfully old-fashioned branch miles from where I lived. They had implicit faith in whatever I did. The overdraft fluctuated violently from one month to the next but all they ever sent me, apart from statements, was the odd communication asking me to join them in congratulating their Mr Smith on his appointment to deputy assistant branch manager of another branch. When I became a dairy farmer, it seemed that I would have to start borrowing serious money to finance the operation and we decided to change to a branch somewhat nearer home.

It was never a totally easy relationship. The bank provides a service like dustmen and drain unblockers but, unlike them, they make enormous profits from their service. They make their living by selling the ordinary citizen money. There is never any risk attached to these transactions as the bank always ensures that they have the power to sell all your goods and chattels and send your family into slavery should you default.

I could never accept the idea that you were supposed to crawl on your belly towards the manager and waggle your bottom violently at him to persuade him to give you the loan of a few pounds. I expected him to come crawling on his belly towards me and plead with me to borrow his money so that his branch could make a good and sufficient contribution to his firm's profits. If my account was ever in profit, they still, outrageously, charged me money to handle cheques instead of paying me interest on it. When I went into debt, they charged me three per cent over the base rate, which meant I was sometimes paying two hundred pounds per thousand for the privilege of a few figures on a computer print-out.

When I changed banks and the manager came out to view the farm, our relationship got off to a decidedly uneasy start when he tried to charge me ten pounds for the visit. It was as if a brush salesman came along and charged ten pounds for putting his foot in your door and trying to flog you a brush. It was not the bank manager's fault; he had just been brainwashed into believing that his was an honourable calling rather than one looked upon with contempt by followers of most of the world's major religions.

Over the course of our relationship, I had to perform many functions that appeared to be perfectly pointless. I was supposed to come up with annual forecasts of future growth and profits, and the cows just did not understand what they were supposed to do. It struck me that it was none of his damn business anyway as long as his money was secured, and when you consider that the calf income could fluctuate by over £5,000 for no foreseeable reason within a year, it seemed to be nothing more than a waste of everybody's time.

The bank drip-fed theories of low finance into my brain which I never appreciated, as all it did was give me something else to worry about. They even gave me a ceiling on my overdraft, which was a scandalous departure from what I had been used to. This was so that I would sweat and bite my finger nails should I ever look like going over it. This was subverted by fixing the top limit several thousand pounds above that which we thought we might reach in even our gloomiest moments.

However, he led me to something which had a profound effect on how I ran the farm. He attempted to shovel great wads of overdraft money at me to buy equipment with which to operate it. Coming in from scratch, we had nothing at all and were expected to buy all the shiny tractors and machines that lie rusting in the hedgerows and barns of most of the country's farms.

I sat down and did my sums. In those days the overdraft was a mere sixteen per cent rather than the twenty per cent it later became. Taking the simple matter of hedge-clipping. We had a good couple of miles of hedges on the farm. Some had been grubbed out by predecessors, but there were still great ramparts of tangled vegetation that needed an annual trim to prevent them advancing triffid-like and taking over the farm. There were two ways that I could tackle these. The bank manager's way was for me to borrow several hundred pounds off him to buy a second-hand machine which I would clip on to the back of the tractor and use to cut the hedges for a couple of days — and then I would have to pass another couple of days picking up the bits so that they would not clog the ditches and get into the cows' hooves.

The other way was to hire somebody to do the job for me. He would come on a dirty great tractor with a couple of thousand pounds'-worth of machinery at the back and do the entire job in a day and a half. And his bill would come to less than the bank's interest charge on the loan to buy my second-hand machine; to that I would have had to add the cost of diesel and repairs. By doing these same sorts of sums in other areas, I consistently found that it would save me money by getting other people in to do almost all the hard work on the farm.

All this lack of hard work in the cause of profitability left me with a great deal of spare time to sit and consider what a hard job farming cows really was. As an outside interest to help fill some of these idle hours, we bought Twinkle. Twinkle was a Gloucester cow which is high on the list of those breeds of cattle that are in danger of extinction. The Rare Breeds Survival Trust was formed to help preserve such breeds so that their genetic material may be retained for some day in the future when its particular characteristics might be needed again. Some of the breeds were in pretty ropey condition when the lifeboats were launched; few more so than the Gloucester.

It is a dairy breed, a rich reddish brown with a white stripe running down its spine and under its belly. It was supposed to be the original supplier of milk for double Gloucester cheese and was almost certainly descended from the same root as the Hereford. By the time that the rescue operation got underway, there were only about forty animals left that could be called Gloucesters with any degree of conviction and many of them had some rather suspect ancestors.

The interesting thing from our point of view was that the breed had already gone into a sharp decline by the time that modern milking and feeding methods were introduced, and so nobody really knew how much milk they were capable of producing or what their characteristics would be under intensive farming conditions. It was fairly certain that we would not become millionaires through their production as there cannot be very many reasons apart from general uselessness why a dairy breed would die out but we thought we would buy one and see for ourselves.

We made a quick tour round the few breeders of Gloucester cattle in the country. Scarcity value was obviously playing a large part in the price, as some of the animals that were offered to us at high figures would have been rejected with a sneer by most of the country's dog food manufacturers. We were even offered, with an air of being done a great favour, an impotent bull which must be one of the more useless of the Creator's inventions. The unfortunate Gloucester had descended in most cases to the ignominious level of becoming a house cow. The average herd size was three: a cow, her calf and the calf of the previous year. We eventually bought our animal, a heifer, at the Rare Breeds Society's annual sale. She was just about ready for the bull and went cheap, we presumed, because she was not classified as purebred, having ancestors that were murkier than even the breed society was prepared to tolerate.

She hitched a lift back to Devon and we put her into the field with the rest of the herd and stood back to watch. Twinkle suffered one grave handicap in her life as a cow in that she did not know that she was supposed to be a cow at all. The bit of paper that came with her gave some clue to the reason. She was the product of an incestuous union between her brother and her mother, which is a kinky enough start for anyone; but the ominous entry on her docket 'reared at the children's farm' conjured up a vision of herds of sticky-fingered brats stuffing her with cream cakes and chocolates, and this was what must have turned her away from the idea that she was a cow.

The rest of the herd took one look as she came into the field and came thundering down the hill to examine this curious brown apparition. Curious brown jobs, in their experience, always had a pair of balls and Twinkle obviously had had none. Twinkle saw this Gadarene rush towards her, looked understandably nervous, moved in behind me and tucked her head under my arm. This was not quite as straightforward as it sounds because her other unique attribute as far as the herd was concerned, was that she had a fine pair of sharp horns.

The herd boss, Number 4, came up and gave her back-end a sniff and bristled when she failed to recognise the smell. The normal procedure expected of a new herd member at this point should have meant that Twinkle untangled herself from my armpit and participated in a series of tedious shoving and pushing matches with every cow to establish her own position in the hierarchy of the herd. Twinkle was too ignorant to observe the usual bovine conventions. She came out from my arm smoothly enough and turned to give 4 a vicious dig in the arse with her horns. After skipping hurriedly out of the way 4 found herself

115

being chased round the field by a furious Twinkle. Every time that 4 stopped to try to turn round and give Twinkle a good thrashing, she received another dig up the arse and was forced to keep going.

The other cows found this absolutely fascinating and stood alongside me with the cud frozen in their mouths like a crowd watching a bullfight. Their mood changed suddenly when Twinkle became a bit knackered and decided to call off the chase to return to the security of my armpit and came bearing down on us. The rest of the herd scattered in panic as she pranced down the field with her horns glinting in the sunlight.

Twinkle really became a bit of a bore for a while. She was convinced that her rightful place should have been curled up in front of our sitting room fire, regarding me as her sole link with her more pampered past. She took to breaking out of her field at all hours of the day or night and standing bellowing outside the house until someone would come to give her comfort. She was not afraid of her new herd-mates but treated them with a profound contempt as being unworthy of her notice. They, on the other hand, found her an object of fascination, rather as we would regard a nuclear reactor: interesting but liable to do something highly alarming or dangerous at any moment.

Since the point of the whole exercise was to integrate her into the herd, we took off her horns. Apart from being liable to do me or the other cows irreparable damage with them as she waved them about with gay abandon, the mangers in the parlour were set so that the animals had to be hornless in order to eat from them. The other cows never realised that she no longer presented much of a threat without them and continued to treat her with profound caution. Twinkle eventually found a soul-mate in Joe. They probably thought that both being brown gave them something in common. Joe took her off to meditate in obscure corners of the farm where he liked to sit and contemplate joys past and joys to come when it was not the nookie season.

These little rambles were stopped when Twinkle showed signs of coming on heat. The preservation of the purity of the race would disappear if she chose as a lover a dirty great Hereford rather than a fellow Aryan. She was locked up for a week while some semen from an extremely well-bred uppercrust Gloucester was imported from the Midlands. She was duly pleasured by the A1 man to good effect. She produced a superb heifer calf nine months later with no difficulty at all, which was more than was expected of her as she never grew to any size. Her udder appeared to be rather ominously small but we put that down to her breed rather than any indication of milking ability. After all, quantity in the human species is not generally a reliable indicator of lactating potential.

It is an unfortunate fact that the friendlier and the sweeter the nature that a cow possesses, the more likely she is to create mayhem in the parlour when she is first milked. Twinkle was lethal. She could not pack the sheer raw power of the Friesian into her kicks but she compensated by having much smaller feet which gave her greater penetration. Altogether it was no great surprise that I could only extract a couple of pints or so from her over her first few milkings.

The vet was summoned to give her a thorough examination when, after a week, her yield showed no improvement. He was fairly scathing. As a milk-producing animal, he considered that she had a rather lower potential than an average nanny goat. Apart from that, she was fine. He was quite right. We disposed of her after about eighteen months, the experiment having proved to be a total failure. It would be unreasonable to claim on the evidence of her alone that the Gloucester breed thoroughly deserves to become extinct, but you will not catch me clamouring in the auction ring to buy another one. She went to the usual graveyard of all such exotics, to a hobby farmer with a couple of funny sheep and pigs where her hopelessly uneconomic nature did not really matter.

Her calf had all the potential that Twinkle lacked. She became part of that terrible group of heifers that escaped the whole time. We named her Venus, the intention being that she would be the first in a line of heavenly bodies. Her mother was always small, but Venus kept pace with her fast-growing Friesian contemporaries all the way through. She kept her horns because of harsh economics. She could not become a sensible profitable member of bovine society with them on but, once removed, her value would have dropped by a hundred pounds. Since we could hardly expect an enormous amount of milk out of her after our experience with her mother, she had to be sold. Horned Gloucesters have a useful zoo potential. Hornless ones lack glamour and do not.

I had been chasing Venus one day, as was my wont after she

and her herd-mates had escaped, when she came to a skidding halt in front of a particularly succulent patch of clover. I took the opportunity to rest awhile before leading her back to captivity and I noticed a milk bottle top had been uncovered as Venus's hoof had peeled back the grass.

Not being an anti-litter fanatic, particularly when it seemed rather appropriate for a dairy farm, I was inclined to let it lie, but my better self won and I picked it up to dispose of at a later date. Cows have the habit of consuming the oddest things and I could see that a milk bottle top might have a rather deleterious effect on the digestion. It turned out not to be a milk bottle top at all and I took it into the dairy to wash it down. In was an Elizabeth I shilling dated 1560. Where it had been lying was the site of an earlier farmhouse most of which had now been covered by the silage shed.

It conjured up a lovely picture of an Elizabethan predecessor chugging up to the local village with a calf in tow to sell it for 1s 1½d in the local market. He wisely decides to invest the odd 1½d in cider at the local pub and reels back to the farm out of his tiny Tudor mind and drops his shilling. His wife must have beaten the hell out of him. Afire with lust for treasure, I bought a metal detector — tax deductible, of course, as its declared use was tracing buried water pipes and checking for bits of wire and other odd bits of metal inside cows which they might have swallowed — and I scoured the farm for my fortune.

I uncovered old horseshoes, ox shoes, nails, ploughshares and even an old iron cart tyre and 22p from the washing green beside the house. I had been told that washing greens were the best place to look since all sorts of goodies fell out of the pockets of trousers when they were hung up to dry. Absolutely nothing of value was found and I made quite sure that the horseshoes were not hung up for good luck. The last time we had done that, we had a very damaging fire within twenty-four hours.

Twinkle was the bolshiest of the cows, but my attempts to make the cows manage themselves produced a pretty bolshie herd. When they had recently arrived on the farm and knew me only slightly, they varied in charm and courage very considerably. Cows like 4, 8 and 23 would have fitted in well at any social gathering, but the herd seemed to split itself into three separate groups. There were the extroverts. They were always first into the parlour, took the lead in duffing up the dog or any deer that might be foolish enough to visit their field and were through any open gates like a flash, to career through the countryside creating as much mayhem as possible.

The next largest group were the morons. Just as people can vary in intelligence and dynamism, so can cows. However, a dozy cow is a very stupid animal indeed. The typical moron was that cow of the popular image, slow, sedate, plodding through life as if her batteries were running down. They could be infuriating because if they were driven, they were quite likely to amble aimiably off in the wrong direction, not through any malice but only because they were not thinking too straight. Morons were always falling into ditches and getting stuck in hedges. The supreme moron was 13; she who sat on her calf and killed it, which was classic moronic behaviour. She was actually a very useful cow because she was the only animal in the herd that would always come when she was called. The name 13 may not be top of the pops in *The Times* birth column but she answered to it and nothing else. It appeared to connect a couple of circuits inside her spongey brain and cause the correct reaction every time. When she came in answer to her name, she often brought the rest of the herd along with her. The cow being a herd animal, once one starts moving in a purposeful manner, the others are quite likely to follow on the assumption that, if they do not, they might be missing something. So whenever the herd was needed, all I had to do was yell '13' and the whole lot were likely to come trailing over the hill.

The other main group were the neurotics. Neurotics were the bane of my life. Those nervy, twitchy cows that can make the farmer's life a misery. That black heifer that I eventually had to get rid of in order to survive was a typical neurotic with psychotic tendencies. They were the animals that at times I would have dearly loved to have assaulted with lengths of lead piping but there was no point because although it would have made me feel better, the neurotic was the one type that did not respond to discipline. I bought three screaming neurotics with the first bunch of heifers. Number 42 was a typical, if rather extreme, example.

She could survive the day quite adequately if nothing out of the ordinary occurred which would suit me fine and I would try to keep the routine exactly the same to suit her. The trouble was that she could always surprise me with her definition of what was abnormal. The time pips on the radio in the milking parlour would be enough to upset her and start her snorting and kicking at anything within range. She was far too neurotic ever to come into the parlour from the collecting yard of her own volition, so I would have to come out of the pit and chase her in. Every day she would indulge in the ritual scurry round the yard. Every day she would run round the yard half a dozen times looking for another exit. Every day she would try to jump the wall and every day she would find that it had not lowered itself from the time before.

I would painfully grit my teeth and pray that she would eventually realise the inevitability of having to finish up in the parlour and cut out all the mucking about. The mucking about was part of her routine of being a cow and deeply ingrained in her character. I changed the routine once. I usually came out to the yard to chase her in wearing the milking apron over the anti-splatter nylon overalls. As it was a hot day, I omitted to put on the overall and such a violent revolution in behaviour was too much for her. She took one panic-stricken look at my lewd bare arms and with one bound was attacking the six-bar gate which was ludicrously beyond her reach.

The fear of naked flesh provided enough adrenalin for her to do the impossible. She cleared it with all but her back foot which slipped between the top and the second bar and there she hung like a slab of beef, with her nose not quite scraping the ground. I waited with mixed feelings for the snap as her bone went since there was no possibility of lifting half a ton of cow out of that position. It was the gate that gave way. The two and a half inch tubular steel bent and she slithered clear, scurrying away up the lane snorting with relief at her narrow escape from the perils of a naked forearm.

She was also a damn nuisance in the fields. The herd would be sitting in the grass, cudding quietly, thinking bumbling thoughts and I would come into the field either alone or with someone else. A few heads would swing incuriously round and look at me with blank eyes. Number 42 would suddenly notice me. She would leap to her feet snorting with alarm and rush round in small circles twanging like a bow-string. The rest of the herd, convinced that the knacker was upon them, would leap to their feet in sympathy and look wildly around for the source of the peril. Some of them would come running towards me for protection. This would

happen every time that I went into the field. Since only a calm contemplative cow makes milk, the havoc that she created cost me good money apart from making me feel like a carrier of leprosy.

My patience ran out one day after the usual scurry round the yard and I telephoned for the lorry to take her to market. I managed to justify the decision by deciding that she went because of low milk yields but, in truth, she went because she was such a complete prick. After she had gone, peace descended. All the cows came in to be milked without persuasion; the yields went up and they sat in the field quite happily while I walked amongst them. In short, I wished that I had got rid of her months before.

I found that the hardest decision to make when I was keeping cows, was when to stop keeping them. Culling cows is an essential part of dairy farming. A dead cow is worthless; it is fed to the local hounds. A live cow sent for slaughter is worth several hundred pounds so that it was very important that as few animals as possible were allowed to die on the premises. When a cow was sick, the gamble lay in deciding if and when she would recover. On a dairy farm, it is uneconomic to get in a hot-shot team of surgeons to do everything to keep the patient alive. If the patient looks as if she might be inclined to turn up her toes, she is bounced smartly off to the knacker.

Some cows can last for fourteen or fifteen years before time catches up with them. Since the national average of a cow's working span within the dairy herd is about four lactations, we are rather short of the ideal. Involuntary culling, as it is called, almost describes itself. That is where there is really no option but to cull because the cow will otherwise be dead or useless. It is in the realms of voluntary culling that life becomes complicated.

Cows vary. My most prolific animal, 10, could be relied on to produce two thousand gallons every lactation. At the other end of the scale was Twinkle who could produce nothing better than four hundred gallons. Number 10 was highly profitable. Twinkle was not. The skill lies in deciding at which point the cow should be disposed of because she is unprofitable. In theory, it should be a perfectly simple business decision. In practice, it was not. Sentiment was involved. Some cows I liked and would tend to hold on to and some I could not bear and was only too pleased to part with. The other problem was knowing exactly when to get rid of what I considered to be an uneconomic cow, particularly in the system that was operated on the farm. The suspect cow calves and she is in the full flush of milk, so you might as well take that before you shuffle off her mortal coil. Her milk slows down a bit but you find that Joe has wasted no time in jumping on her and she is well in calf. That you might as well take, since it is worth the best part of £100 and so you wait until she gives birth and then she is in the full flush of milk again.

Another problem is that you are never quite sure about the quality of the beast that will replace your cull. You can reckon on a couple of hundred pounds difference between the value of the departed and the cost of the replacement, so you are already demanding that the new cow will work fairly hard to repay that. If you run an average herd, it is not difficult to calculate that the chances are even that your replacement will be above or below average. If you keep a herd that is slightly better than average as, we thought, was ours, then it becomes more difficult to buy in a cow that would be up to scratch. For these reasons, our cows had to work pretty hard to earn a ticket to the meat factory or the market.

Apart from Twinkle, we disposed of five cows over the years on the grounds of poor yields. Admittedly, in the case of the loony, bantam-bashing heifer and 42, it was stretching the truth a bit to call it poor yields rather than my own dislike of them. I once had a quick flash of independence and bought a cow entirely off my own bat without asking for advice from the experts. She was a very elegant animal with an ancestry that included most of the Friesian aristocrats over the last thirty years. I was very pleased with myself as I bought her comparatively cheaply. She duly calved down and came into milk. For a week or three, she and Twinkle vied with each other as to which could produce the least amount. I had the vet over to check her out. He was very impressed with her and said that she would stand a good chance of winning half the cattle shows in the country and was in perfect health. As a milking cow, he cheerfully said, she was useless.

There are three ways of getting rid of a cow. Straight to the meat factory where you throw yourself entirely on their honesty as regards price. Into the barren ring at the local market whch is the usual place for old screws to end up. The final and financially most attractive way is by putting them through the market's freshly calved ring. Here go all the cows and heifers that have recently calved and their ultimate destination is somebody else's herd where they are meant to turn into productive cows.

It was hardly moral to put an animal of such intrinsic uselessness as my pedigree failure into this ring, and so I gave the haulier strict instructions not to tell me where she ended up but to use his own judgment on where she would fetch the highest price. This salved my conscience and my pocket and I kept well away.

Another animal that I disposed of through this method was 18. She was that incredibly slow milker with teats that erupted from her bag at all sorts of unexpected angles. She was a dear old soul but I would go up to the house for a leisurely breakfast after I had put the cluster on her and all the other cows would start playing patience or sardines to kill time until she was finished. There was nothing much wrong with her yield but the speed at which she milked was just too much for my patience and that of the cows.

The technique, in these freshly calved markets, is for the vendor to enter the ring with his cow to show that he has confidence in her and is prepared to stand by her. I was prepared to do this with 18 although I had not been with the pedigree disaster. The other thing that the vendor is supposed to do is to walk round the ring with her, tugging at her titties at intervals so that the milk gushes forth and thus demonstrates the cow's good nature and the ease with which she can be milked. To make the milk gush forth from 18, you either needed hands with the strength and size of a gorilla's or else a pump with the power to drain a coal mine. My peers stood and watched with a jaundiced eye as I swung from each teat in turn, producing a couple of tiny drops of milk from one and damn all from the others. Number 18 behaved very well considering that she was not used to such abuse but she became annoyed eventually and walked off after knocking off my farmers' market trilby and treading it firmly into a pile of muck left by the cow before.

I got eight freshly calved heifers through this market once. I did not trust my own judgment, obviously, but my friendly auctioneer was doing the selling and so I gave him the commission to buy on my behalf those that he thought would be suitable. I stood and watched beside one of those gnarled old farmers who have seen it all, and who infest all markets.

'He's buying for somebody,' said the old man. 'Just you watch. He's taking bids off the bloody walls.'

'Is he really?' I replied. 'How very interesting.'

'Stupid bastard.'

'Who?'

'The fellow he's buying for, of course. You can never trust these boogers.' The old boy was really cheering me up.

'Perhaps the fellow he's buying for doesn't know very much about cows. After all, the auctioneer must know where all these animals are coming from and have a much better idea of their potential than anyone else.' I felt quite proud of myself.

'He'll find out plenty about cows when he tries to milk some of those screws.' A great gangling heifer with a rather mean look in her eye cantered into the ring with its vendor in attendance. The audience leaned forward in anticipation as the owner cautiously approached her and bent towards her udder. 'You watch this,' said my neighbour. I watched the heifer let go with an immaculate hoof and the vendor doubled up on the straw-covered floor of the ring clutching his knee. The crowd let out a collective sigh. 'I pity the poor fool who ends up wi' 'ee.'

'So do I,' I replied fervently and sent out Spock-like thought waves to ensure that the auctioneer got the message. The vendor was now hobbling round the ring with his heifer in pursuit.

'A fine spirited animal,' said the auctioneer. The bidding got under way. I caught the auctioneer's eye and shook my head vigorously. He beamed back at me. I shut my eyes and prayed.

'There aint no use in praying, lad. The booger's bought it for you.' He damn well had, too. He did buy me some quite extraordinary animals on that day, the bull's beloved being one of them. They did surprisingly well although most of them had eccentricities. About the only one that never lifted a hoof at me was the great gangling heifer. She turned out to be a honey.

# Chapter Eight

Involuntary culling was almost always because of ill health. One of the most common reasons for cows being culled is because they cannot be got into calf. On our farm, not one cow went to heaven for that reason which is a tribute to the power and potency of Joe. Three cows actually died on us including Number 6 who was squashed by the falling roof after the blizzard. It is very annoying when an animal dies because you always have the feeling that if you had been a little more observant or taken a little more care, you could have avoided it.

The first demise was due to my rank inexperience. An animal calved, became infected internally and died. I had had the vet in to assist with the calving and he had decided not to protect her with an antibiotic after rooting around in her vitals. A perfectly reasonable professional judgment on his part, had he been dealing with an average farmer who would have realised that his beast was in trouble a couple of days later and would have called the vet back. I only noticed something amiss an hour before she died and the vet turned up just in time to hear the death rattle. It is an expensive way to buy experience.

The most upsetting cull was 23. She had always been the largest and one of the sweetest-natured animals. She went off her food and looked a bit sorry for herself about a week before she was due to calve. I shoved her into the sanatorium paddock outside the parlour and called for help, thinking that there might be some minor upset due to her pregnancy. The vet's diagnosis was no comfort and I got opinions from both partners in the practice to confirm it. She had swallowed a piece of wire at some stage in her life which had now shifted and formed an internal abscess for which there was no remedy. I held on to her for several days, pumping her full of antibiotics, but it became obvious that she would eventually die.

To give her her due, she tried to save as much from her débâcle as possible. Two days before the lorry was due to come for her, she ejected her calf. Keeping in with the traditions of the farm, she did not drop it in a daisy-covered arbour, but six feet down, straight into a bramble-filled ditch, and then she started to bellow. I faithfully turned up in answer to the bellow and with considerable difficulty scrambled down into the brambles at the bottom of the ditch. Inevitably, it was one of those wriggling hundredweight calves rather than a nice placid little one and I had to engineer a straight lift to eye level to get it over the muddy edge of the bank. It should not have been too difficult, if a little messy. Unfortunately, I was dealing with a frantic mother as well as the calf. As hard as I was pushing the calf up, she was pushing it down again with frantic licks of her tongue, half of which were slapping me in the face rather than her correct target. I had to haul the calf out backwards, presenting the mother with my backside which she preferred not to lick.

Of all the outside influences of the farm, the vet probably yielded the greatest clout and extracted huge sums of money from me each year in fees. There is a marvellous difference between a vet and a doctor. Ring up the vet in the middle of the night and he will come running with his coat over his pyjamas. Call a doctor and he will tell you to take two aspirins and come to see him in the morning. The difference is explained by the fact that the vet sends you a bill and the doctor does not.

The second time that the vet came to the farm was a mixed success. It was during one of my very first milkings and, about 5.30 am, I was about to shove on the cluster and brace myself for the inevitable kick, when the cow gave an agonised groan and sat down. This was not normal behaviour. The book said nothing about cows sitting down in the parlour when they were being milked. The other cows found it rather abnormal as well and lifted their tails and gave little limbering-up twitches of the hooves to warn me that they were about to indulge in mass hysteria. I hurriedly got out of the pit and with a mixture of kicks and prods persuaded the cow to crawl towards the end of the parlour where I could examine her. A cursory glance showed that she had a huge swelling on her left side below the rib cage and I fled up to the house to call the vet.

'Ah', he said. 'It's obviously bloat.' I sounded blank. 'Bloat is the same as wind in a baby.' I had visions of having to put the cow over my knee and pat her on the back.

'And what are you supposed to do about bloat?' I asked.

'Drench her or stab her. Hmm. It could always be frothy bloat I suppose.' It was beginning to sound worse and worse.

'Why don't you come out and have a look at her?' I suggested. So he came. He sang loudly in the dairy while donning his protective waterproofs. The cows still waiting to be milked scuttled round the yard in panic. He was not very tuneful. He came into the parlour and looked down at the cow, now stretched out and looking as if she was about to peg out at any moment.

'Shall I stab her?' he asked.

'For Christ's sake, don't ask me,' I replied. 'I'm supposed to pay you to make that sort of decision.'

'But do you think I ought to?' I was beginning to panic a bit. It was a bit too reminiscent of the time my wife had a baby and the nurse in attendance asked me if I could tell her how she was supposed to operate the anaesthetic machine.

'No. Perhaps I won't stab her just yet.' He stuffed a tube down her nose and we both listened to the Rice Krispie-like noise that was coming from her innards. Apparently that was a sign of frothy bloat.

'Do you have any Fairy Liquid?' he asked.

'Well, not on me.'

'Could you get some, please?' I went up to the house to get some Fairy Liquid wondering if I had the right man to the job. When I got back, he was serenading the patient with 'Oh what a Beautiful Morning.' He shook up a wine bottle containing water and a liberal dash of Fairy Liquid and poured it down his tube. The cow emitted a thunderous, window-rattling belch and got to her feet and ambled out of the parlour. He went away singing happily but, as far as I was concerned, he could probably walk on water or raise the dead. Bloat apparently came about when there was lots of clover about, the patient had been particularly greedy or, in the opinion of a neighbour, she had been eating while facing into the wind. I never got very much of it, which was a little disappointing as it is one of the truly dramatic moments of agriculture when one is faced with an apparently moribund cow with bloat. One sticks a

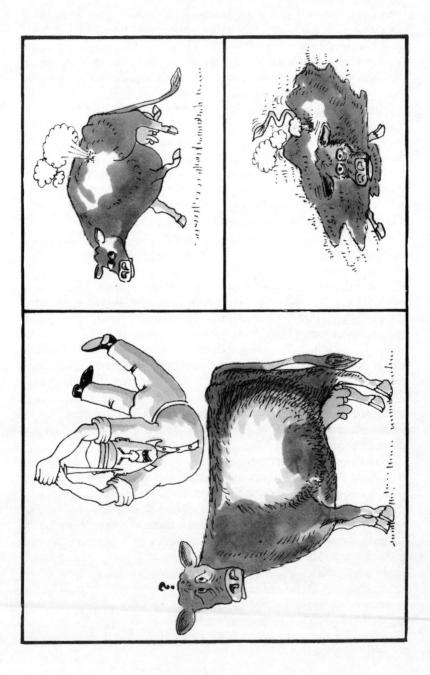

bread knife into its guts with a screech of 'Banzai' and the animal promptly gets up and walks away with the knife wielder haring after her to retrieve his weapon. I had the joy of doing it once in an emergency. I also had a few calves that became bloated and I developed a neat line in sticking them with hypodermic needles so that the excess gas hissed out.

The prime disease of all dairy cattle is mastitis and ours were not immune. The disease is a real nuisance because it is so difficult to avoid. It is caused by bacteria entering the udder through the tit and setting up an infection which can either show itself as nothing more than a few clots in the milk or can run the gamut from swollen udder right up to death. Our cows were regularly afflicted by a particularly pernicious variety — summer mastitis. This usually hits the cow when she is dry and waiting to calve but it can attack heifers just as easily. In every case that we had it, the cow or heifer in question lost the affected quarter and, once, the cow died.

That was particularly frustrating as one is supposed to contain or even cure the disease provided that it can be caught at an early enough stage. The dry cows were checked at twelve-hour intervals but, between one check and another, the bug struck. Antibiotics were poured into the afflicted cow immediately but she was clearly very ill. The vets kept her alive for a week on a mixture of intravenous feeding and antibiotics until she turned up her toes. That showed another very good reason for culling a sick cow if there seems to be any doubt on the prognosis of her case. Apart from losing the value of the cow, I had a vet's bill for eighty pounds to pick up.

Usually summer mastitis would not be quite so devastating in its effects. I would end up with a cow that would lose all milk in one or two quarters. For the first couple of years the disease only struck cows that were such poor yielders that I was thinking of getting rid of them anyway so I could palm the responsibility of turning them into meat pies off on to the bugs. Latterly it started to hit some of the better cows in the herd and these we kept, and took what milk they were prepared to give from their three remaining quarters. The bacteria causing the disease are supposed to be fly-borne and there was no doubt that the flies were the source of considerable irritation to the herd in high summer, and any wound or sore would have the flies clustering round in a way that must have driven the wounded potty. If they became particularly bad, the herd would clump together and face inwards and swish their tails furiously to drive them off.

The entomological equivalent of the B52 bomber is the warble fly, which strikes terror into cows. It has the nasty habit of laying its eggs on the cows' backs so that the hatching maggots burrow under the skin and create havoc before emerging the next year. The insect is on a losing wicket since it is comparatively easy to kill when on the cow and it will probably soon be eradicated. But when the adult fly is cruising around looking for a good laying spot, the cows go crazy and sprint all over the field with their tails in the air in their attempts to avoid being the chosen one.

In summer, I did my best to help the cows by spraying them with a long-lasting insecticide which was supposed to kill flies for a fortnight before it needed to be renewed. Part of the idea was that by killing all the little breeders at the beginning of the season, the total fly population would be considerably reduced. The cows thought that the results were marvellous, and for a few days after the spraying there would be peace and tranquillity. The flies could be seen actually dropping dead as they landed on the cows until the effects began to wear off and the population of insects would build up again.

The herd were all for the effects of having been sprayed, but they took a very dim view of the actual spraying. The blurb on the tin said that the cows would rather enjoy the nice cool shower-bath when they were sprayed from the knapsack sprayer, but our bunch had to be chased round and round the yard and tried to hide under each other leaving the weakest on the top of the pile. Joe even shook his head at me in a threatening fashion until my look of hurt surprise caused him to slope shamefacedly up the yard.

Nothing that we tried ever stopped summer mastitis and nothing that we tried ever really worked against the normal kind of mastitis. The Milk Marketing Board took to calculating the levels of mastitis in everybody's herd and issued dire guidelines on its economic consequences. We were all too aware of the consequences, but nothing we did seemed to improve matters very much. The most immediate result of an attack of mastitis was that we were supposed to discard all the milk when the animal was being treated. A good cow would be giving around eight gallons of milk a day in the full flush of production after calving. She gets a clot or two in one quarter. The rule book states that thou shalt discard the milk for the three days that the quarter is being treated and keep discarding all her milk for another three days afterwards. That amounts to fifty-six gallons of the stuff being poured down the drain, and then there is the bill for the drugs on top of that.

If you were naughty and failed to discard all the milk from all those quarters, the risk was that you could face the full wrath of the MMB who would inflict on you assorted tortures for selling them contaminated milk. We all live in a risky world. It is a very dangerous game as the Board can refuse to buy your milk if they catch you at it more than once in a six-month period.

It is quite understandable, quite apart from poisoning the consumer which cannot be very good for business. I once visited the milk factory which took my produce. They turned this precious lifeblood of my existence into a quite revolting rubbery cheese which they made by the ton before sticking it into plastic packets. If antibiotics get into the milk that they receive, then it knocks off all the quite revoltingly rubbery cheese-producing bugs, and although it might be excellent from the point of view of the nation's palate, it gets a bit expensive if each discarded batch weighs a ton.

The frustration of our regular fortnightly case of clinical mastitis, often with a sick cow and swollen quarter, was compounded because we faithfully carried out all the recommended preventatives. We dipped the cows' teats in disinfectant. We tried to protect them during the dry period with long-lasting antibiotics. The water used for washing them was carefully disinfected and we scrubbed them down with paper towels before the clusters were put on.

There was also the worry about the quantity and frequency of the antibiotics that had to be used. They were the only way of treating the disease and it was all too easy to have a few bugs left behind after the course had been completed, which could multiply and create a strain resistant to that particular drug. Unlike the sneaky variety which, I believe, lives in the sewers of Bristol and falls on all known antibiotics with shrieks of glee and munches them down, our bugs seemed to be geared to the average intellectual level of the rest of the farm and forgot after a couple of months of non-use that they were supposed to be resistant to a particular brand.

We even spent pounds on an expert to tell us how to beat the mastitis problem and he grew gloomier and gloomier as the months passed, during which we religiously followed all his instructions and the levels of the disease continued to rise. We sacked him in the end because we felt that we knew as much about it as he did, his brain having been sucked dry. We ignored his ultimate weapon which was to knock off any cows that contracted mastitis more than a certain number of times. This seemed to be the height of ingratitude. The poor brutes were working their titties ragged on my behalf and I was supposed to slaughter some of my most charming animals in order to satisfy a theory.

We had other diseases and epidemics. There was an outbreak of New Forest Disease which affected virtually every cow in the herd. This was another of these mysterious infections — probably fly-borne again but our particular attack occurred in mid-winter when there are not that many flies about. New Forest Disease attacks the eye and is obviously extremely painful to the sufferer who is temporarily blinded and whose eyes can be permanently damaged. None of ours permanently lost their sight but it was a pretty shambolic month while it was at full blast. We would have anything up to a dozen totally blind cows at any one time stumbling about in the yard with violently watering eyes. They crashed into walls and fences and one was spooked by a cat right through the barbed wire fence into the slurry pit. Being blind, she did not come out the easy way but spent half an hour undulating her way across to the other side where I put a halter on her and led her back.

The treatment was not much fun either. Tubes of ointment squirted into the eye had no effect and, instead, the vet had to inject a couple of ccs of antibiotic into the inner eyelid of each cow every couple of days. It is no easy matter to hold the head of the cow still enough for long enough to allow a needle to be jabbed in beside its eyeball. It took enormous strength on my part and considerable skill on his. The success of it was that every cow recovered completely.

It is in the realms of doctoring that single-handed dairy farming shows its limitations. The cows were considerably bigger and stronger than me and persuading them to co-operate in anything

133

that required either the knife or the needle could lead to problems. We had a crush attached to the bull pen. This was originally built to contain a cow and trap her head between a couple of iron bars, at which point a bull would be released and leap upon his quiescent victim. Why any bull should need to have his women tied down, I cannot imagine.

By using the crush, I could trap a cow and examine her and doctor her to my heart's content — within reason. Although the head might have been trapped the back end could still do very much as it wanted. The other problem was actually persuading the brutes to put their heads into the desired position so that I could trap them between the bars. The technique called for was to put the cow in the bull pen, not always plain sailing in itself, and then drive the animal into the crush passage until her head was through the bars. Then she was politely asked to stay where she was for a few seconds while I nipped out of the pen round to her front and moved the bars together. Sometimes a particularly stupid cow would co-operate with this performance the first time you tried it on, but the next time not even the dimmest cow would really help.

Subtle methods of persuasion were used to assist the cow to the appointed place: the bucket of cake or the wisp of hay on the other side of the bars, or the stout length of swishy cane to beat their bums until they went forward. The only sure way was to grab their noses and pull them through but even then you had to let them go in order to shut the bars.

The most usual reason for wanting an animal in the crush was that there was something amiss with her feet. Cows' feet are always getting bruised, or getting grit in them, or just going rotten. Trap your cow, pick up her hoof and hack away at it until you have sculpted it to your satisfaction or removed the obstruction, and then you often have to finish with a milk-withholding shot of penicillin in her bum.

The average cow had four methods of dealing with attempts to doctor her feet. The first was the simple kick aimed at the groin. The second was the shake and it was grim work trying to hold on with your teeth rattling and your eyeballs vibrating in their sockets. Then she could always lean her full weight upon you as you supported the foot; and finally, if it was a back foot, she would shit on your head and that rarely failed to create a distraction. It was always a brutal, bruising business with all the animals except Joe. I could always walk up to him in the field and pick up a hoof and have a look and even stick a needle in his behind, although he got a bit huffy if it happened too often.

The monthly vet's clinic was the one time that my order rather than their order had to be imposed on all the cows. The whole herd would be shoved into one half of the cubicle shed and then run through the crush where the vet would check them for illnesses or stick a hand up their backside in a mighty goose to check on the state of their pregnancy. He would, with a certain amount of groaning and complaining, hack out the odd hoof and I would post myself at her head where I would mutter obscenities into the patient's ear to provide a distraction and stop her kicking the vet to death.

Sometimes we would have a veterinary visit sponsored by the Ministry of Agriculture. These were to run TB and Brucellosis tests on the whole herd. On these occasions the visiting fee would be paid for by the Ministry, and so we would load on as much other business as we could. Brucellosis, apart from Foot and Mouth, is probably the most devastating disease that can hit a dairy herd. Its symptom is abortion and is highly infectious. Once the cow had aborted, she would be out of production for nine months until she could have another calf and by that time she would have infected every other cow in the herd. We had the disease all around us at one point, but our herd managed to keep clear. It has now been almost eradicated throughout the country, but the Ministry still keep a very close eye on the situation. They insist that the farm has an isolation pen so that any animal that aborts may be isolated from the rest of the herd while the experts check on whether or not it is Brucellosis.

The most contagious item is the afterbirth and, as all our cows calved in the field with the rest of the herd, isolation seemed to be a case of locking the stable door after the horse had bolted; particularly when I had two cows abort in one night and I found the rest of the herd playing tug of war with bits of their afterbirth. I rescued the bits that had not been consumed and the vet reverently laid them to rest in the boot of his car and whisked them away for tests. I locked the patients up in their isolation pen, where they behaved themselves quite well until milking time when they both leapt the gate and joined the rest. Sod it, was my reaction, and I decided that the whole herd should be deemed to be isolated rather than just the two of them. False alarm was the vet's considered judgment but several weeks later, long after I had forgotten about the incident, I recieved a letter from the Ministry telling me that the two had not got Brucellosis and that they could now join the rest of the herd.

Cows are either in the full bloom of health or look as if they might be on the point of death. One of those tricky decisions that I never really knew how to take was determining at what point the vet ought to be called in. Every time that he visited there would be a steep bill to meet, so you would have to consider carefully whether his presence could be justified. However, your cow cost £500 and that could pay for quite a few vet's bills and you certainly did not want to lose her. Every time that I did not call in the vet, I doctored her myself, which meant poking and prodding her and sticking in a syringeful of whatever I had around. Then I would chew my nails and wait until she died or recovered. When I called in the vet, he would poke her and prod her and stick in a syringeful of whatever he had in his boot. I was paying my visit fee so that he would have to bite his nails rather than me bite mine.

Being a farm vet cannot be easy because your paymasters reckon that they could do just about as good a job as you can. You also tend to be called in to clear up the results of the farmer's own attempts at a cure. The vet's ultimate power lies in the drug bottle. This is where the dark mystery of his profession is at its murkiest. Drugs are grudgingly doled out in exchange for large sums of money, and very few farmers can differentiate from one immensely complicated pharmaceutical name and another, so the vet can use them to preserve his professional secrecy. Vets have further powers in that a variety of government regulations forbid the doling out of the vast majority of antibiotics to the farmer, mainly to keep them out of the jaws of those ravenous bugs in the sewers of Bristol. If a visit costs £10, and all that is needed is a shot in the bum for one cow that any one-eyed, half-witted junkie could administer, no wonder the farmers smile and cringe before their vets in the hope that they will leave the subsequent doses on the farm and thus save them each £30.

One of the more dramatic health upsets that the cows enjoyed was milk fever. This is caused by a shortage of available calcium in the blood stream and usually occurs round about calving.

About half our animals would go down with milk fever on calving. You know an animal has got it when she quietly keels over and dies within about twelve hours. The cure is as equally dramatic as the symptoms. Get hold of a dirty great needle, jab it into the cow and slosh in a couple of pints of calcium. The cow should then blink a few times, get up and walk away. The expert does not just stuff his needle in any old where. He sticks it into the jugular vein. This is high-powered stuff that can be learned by studying the way that the vet does it.

I had to learn this method. Since very few of our calvings were superintended, a cow could have been wallowing and groaning for a few hours before I came upon her. A case in point was 17, a sound cow who poured out milk like the Chatsworth fountain during her first lactation and never quite lived up to her promise afterwards. Number 17 was in a fair degree of trouble when I came upon her, lying on her back with all four legs in the air and her belly distended. Frequently, when I would be busy doing something else, I would casually look into the field and see amid all the other quietly cudding cows one sprawled out on her side motionless. Not another, I would say and ignore her. I would keep sneaking glances at her and after ten minutes immobility, my nerve would give out and I would go to see if she was all right. At 200 yards, I would begin to sweat. By 100 yards, I would be pretty sure that she was dead. At 20 yards, I would be wondering how I could afford to replace her and how I could get her out of the field. At six feet, she would roll over and give me a look of mild enquiry.

Number 17 was as dead as any cow that I had seen. In fact I did not bother to examine her but checked over her calf instead. When I got around to its mother, she was still alive but emitting I-am-about-to-die sighs instead of breathing. In a situation where milk fever has taken as strong a hold as that, it is useless to put the calcium under the skin of the rib cage, which is the norm, since the patient's circulation will have deteriorated to such an extent that it will not distribute the liquid through the system. That is when you need to be able to find the jugular in a hurry. You have to dig around in her neck until a fountain of blood indicates that you have hit the jugular and then you pour in the calcium as fast as you can without giving the victim heart failure.

If a woozy cow, swaying gently on her feet, is a prelude to milk fever, then an animal strutting towards you with a manic glint in its eye would very likely be a prelude to the other common deficiency, which is magnesium. This was a much more alarming happening than milk fever as the usual first symptom was a corpse lying in the middle of the field. The animal becomes more and more excitable and nervy until its heart eventually gives out, and the whole progress can take place within a couple of hours. To make matters worse, if you happen upon a cow before she has pegged out, the act of inserting the needle so that you can administer the magnesium could very easily give her sufficient shock to kill her.

The first year that I kept cows I never saw the deficiency and had hardly ever heard of it, and so I rested easily. Next year I had the frighteners put on me by a couple of reps and spent hundreds of pounds on a constantly available supply of magnesium to prevent it. Then I decided that I was being a bit like the man who kept clicking his fingers to keep the elephants away, and so I stopped buying it. Then I got my one and only case.

Typically for this farm, the victim made a real meal of it. I brought the victim in to be milked as usual and she started to leap about and loose off wildly inaccurate kicks at me. Since she was normally a placid cow, this was rather out of character. I went to put on the cluster and touched her teats. She let out a noise like an air-raid siren and leapt on to the back of the cow in front, rather destroying the peace and tranquillity of the milking. She then started to kick with all four legs at once and toppled over the rail down about six feet into the pit. Avoiding cows that come crashing down towards you from a great height is one aspect of dairy farming that you cannot pick up from a book. You have to learn through experience. I sidestepped smartly and she landed at my feet with a thud that shook the entire building.

As usual, on winged feet, I rushed to summon the vet. The cow had recently calved and I was fixated on milk fever although this was unlike any milk fever that I had ever seen. While waiting for him to arrive, I continued with milking. The cow was lying on the floor semi-comatose and groaning. The pit was a pretty tight fit when it contained no one but myself, but sharing it with several cubic yards of cow put a considerable strain on its facilities and I had to clamber on top of her to reach some of the more inaccessible levers. Every other cow that came into the parlour froze with horror when they saw their herd-mate lying in the pit.

This was a departure from the norm with a vengeance and they showed their disapproval, every one of them, by lifting their tails and spraying the parlour artifacts, myself and the comatose cow with the appalling results of a diet of lush young grass.

On his arrival the vet carefully hosed down his patient so that he could examine her and covered all his options by deciding that she was suffering from a deficiency of both calcium and magnesium and sloshed in a gallon or so of each followed by an aspirin to help the headache caused by her fall into the pit. He then rushed on to his next emergency call, leaving me to clear up. I thankfully watched the last panic-stricken cow scurry out of the parlour and prodded the recumbent patient who grunted back at me. I dismantled the railings at the end of the parlour that prevented the cows from coming into the pit, in the hope that it might make her exit more easy. I prodded her hopefully and she rolled over and stuck her legs in the air. I thought a rope might be useful in case she needed to be winched out and I climbed out of the other end of the pit to get one from the dairy. The patient, brown, steaming and reeking, crawled up the steps after me. OK, dear, if that is the way you want to do it. You can now turn and go through the mangers and out the door at the far end. The cow turned and squeezed through the gap in the railings and went back down the stairs into the pit and lay down again and groaned some more. I stood torn between weeping with frustration and foaming at the mouth. She could have walked out the door quite easily. I went down and poked her again. She appeared to be unconscious once more. The great light dawned. I used the ultimate weapon. Savagely, I turned the radio over to Terry Wogan. She jumped to her feet in horror, fled up the steps and out the door.

We had other odd illnesses and diseases. There was a run of cows with excessively high temperatures for no veterinarily indentifiable reason. They all recovered but it played hell with the milk yields. I used to get through several thermometers a week. The technique of measuring the cow's temperature was that you stuffed the thermometer up her arse. I would do this during milking and leave it for a minute or two while getting on with the job. There would frequently be the pitter patter of falling dung, in the middle of which would be the thermometer. The pat would be carefully dissected and, as often as not, the instrument would be shattered. Even more disconcerting were the occasions when the thing disappeared inwards rather than outwards. I would sometimes manage to grab the end of it before it disappeared and an undignified tug of war would take place, with the cow frequently being the winner.

One animal started to smell like a week-old corpse, and as she seemed to be quite happy, I left it for a few days before cornering her for examination. She appeared to have bashed her back against something and there was a wound all along her spine with bits of loose vertebrae floating about in it. I cleaned the bits out and she was fine. Cows show remarkable resistance to mechanical injury and pain, which can be a bit of a bore if you are behind them with a stick.

Some other animals were in the collecting yard when the vet came to wash out the guts of a cow that had failed to cleanse properly after calving. Being a vet must at times be a bit like being the reverse of Francis of Assisi. Your job is to cure animals, but all the dumb brutes know is that every time you come near, you chivvy them around and usually stick needles or hands into sensitive portions of their anatomy. The sight of the vet will send every self-respecting beast scurrying desperately for the sanctuary of the woods. The cow to be washed out was cornered in the milking parlour, while those left out in the yard tried to hide behind water troughs or scale walls just in case the vet decided to pick on them.

He went and I cleared out the cows in the yard. Number 15 had a very bad limp. I thought that she must have wrenched her shoulder as I had recently had a very similar case when a cow had tried to mount Joe rather than the other way round and he had taken violent exception to the perversion. Just in case, I called the vet back. Vets sometimes came and went like yoyos. Number 15 had broken her leg. The vet was stricken with remorse, because if he had not come she would not have been in the yard, and therefore she would not have broken her leg. I did not quite follow his reasoning. If I had not written this book, you would not be reading it and you could instead be making a million by speculating in lambs' trotters in the Chicago futures market. I am very sorry.

The vet told me an even more awful story. One of his customers had bought an extremely smart and expensive young bull and called him in ostensibly to check it over, but in reality just to show it off. The vet was placed in the yard while the proud owner went to get the bull. It pranced into the yard, tripped and fell down at the vet's feet and broke its neck.

The vet's advice on 15 was to send her to heaven, but since she had only just had her first calf and stood to lose me a couple of hundred pounds I decided to hang on to her for a few weeks to see if she might mend. We stuck her in the bull pen on a thick carpet of straw to prevent her jarring her leg and threw in a couple of calves to turn the milk that she was giving into some sort of profit. At the end of that time, we opened the gate to see what had happened. Out shot two gargantuan calves as fat as butter and after them, like Long John Silver on skates, hurtled 15. I gave her a hearty thump on her damaged leg but she did not flinch and so I deemed her cured. Her leg was never really right as it hung at a rather awkward angle. To have plastered it would have been impossible since the break was too high up, but the amount of milk that she gave after she came out of the bull pen was greater than what she gave before going in. Over the years her limp gradually faded and her gait, from resembling a galleon swaying through the Caribbean after a hurricane, softened to the same in a stiff breeze. She still looked terrible but remained a highly profitable animal.

Our herd was what was known as 'commercial' rather than 'pedigree'. Commercial cows are designed to produce milk. Pedigree cows are designed to produce milk and pedigree calves which, hopefully, can be sold by their owners to other breeders for vast sums of money. With Joe as the husband to all our cows, going pedigree would have been a bit of a waste of time since all our calves were Hereford–Friesian crosses. However, I had a brush with the smart end of the trade when I bought twelve pedigree heifers off a neighbour.

They were all in calf to a rather smart bull and he wanted to buy back the calves from me. If I just let the heifers calve and gave him the calves, they would have been useless to him. Their father was a smart pedigree bull and their mother was a smart pedigree heifer, but they would be common or garden little mongrels unless I went to the expense of registering myself as a pedigree herd owner. This process mysteriously improved the value of the calves by fifty per cent. A pedigree cow also has an official name rather than merely a number. Not Blossom or Daisy but, to take a couple of my own that had been pedigree before descending the social scale to ourselves, names like Loxhay Bertha 15 and Briwood Reflector Modern. Even the Gloucester was blessed with the name of Twinkle Hopeful Venus.

The calf buyer paid me to register as a pedigree breeder — a bore of enormous proportions as it involved me with a completely new set of bureaucrats who demanded meticulous form filling so that the calves could be accurately registered. In revenge I chose the herd name of Scrub so that all those smart little calves would have to carry Scrub in their names until the day they died. They were all ultimately sold to France because the French were the only ones who did not know what the name meant.

When I joined the Breed Society, I was inundated with a load of leaflets all describing the perfect cow. The only criterion I used was whether or not she produced lots of milk but that was the one aspect that the leaflets ignored. She should not have any black on her feet. She should have fine skin, have well sprung ribs and look feminine. My dictionary defined feminine as 'womanly'. I could not see a 38-22-36 cow winning many shows.

In our own herd, we searched long and hard for a common factor in the cows that yielded most. My best cow had one quarter that was twice the size of her others so she was a dog in the eyes of the experts. Number 8 had chronically bad feet and groaned every time she put them down. Number 32 was too small; 33 had pointed tits and a bow-shaped back. The only possible factor that I managed to isolate was that all the really productive cows had high-pitched moos. But, when you think about it, that did not seem any worse a basis for selection than the length and thickness of her eyelashes.

# Chapter Nine

The ideal towards which I constantly strove was that image of the immaculate dairy farm. Contented cows in rolling green meadows with just the odd buttercup scattered around to break the monotony and provide colour contrast. The grass should always be precisely four inches high. All buildings and fences should be faultless and probably painted white and, for perfection, there should be a full choir from the Women's Institute singing 'And Did Those Feet?' somewhere in the middle distance. Our land sprouted all sorts of excrescences that spoiled its symmetry and, against these, I waged implacable war as they offended my sense of artistry and beauty.

The worst of these pustules on the fair face of the farm were the docks. On the grazing half, the cows tended to eat them off, but on the silage side they multiplied in sinister profusion. Every few weeks they would be cut along with the rest of the grass but they would fight back and proliferate in revenge, creating clumps of livid green that towered over the surrounding grass.

In the beginning, I bought a little scythe and during idle moments, of which my agricultural system gave me quite a few, I would be off into the fields, hacking away at the hated enemy. There were too many for this form of attack, so I tried chemical warfare by spraying them with killer chemicals from a watering can. For about ten days from dawn to dusk, I was out pouring on the poison, chuckling like Bluebeard at their agony. The docks, being even stupider than the cows, failed to understand the point of the exercise and treated the weedkiller like fertiliser and grew even stronger. I had a greater degree of success with biological warfare. I found a fat green beetle in a remote corner of the farm that appeared to specialise in eating docks. These I scattered over the fields and they seemed to keep the worst of the docks in check.

The cancer of the grazing half of the farm were the thistles. We had three different sorts: spear, marsh and creeping. The spear thistles were easy as they were large and timber-like and fell with satisfactory finality. Marsh were more deceptive. If anything, they were even larger than spear thistles but they popped up every other year. I would think that I had finally killed them off but they would merely be biding their biennial time.

The creeping thistles were the worst because keeping them under control was a never-ending struggle rather like having to paint the Forth Bridge. We had a large acreage next to the farm that had been let as seasonal grazing over the past twenty years. The owner spent no money on it, and the tenants were only interested in making sure that they extracted the last penny's worth of their rent each year. This became an agricultural Chamber of Horrors as the hedges went to rack and ruin, and successive waves of weeds including creeping thistles stole over the boundary and attempted to occupy the farm.

When was a weed not a weed but a precious wild flower? There could be no doubt with thistles because as the cows refused to graze amongst them and get their noses pricked, they led to considerable amounts of wasted land. Nettles could erupt in great clumps in the middle of a field. Nettles were eaten by caterpillars which turned into desirable butterflies. There was a weird form of perpetual motion going on there. You had to encourage the butterflies so that their caterpillars would eat the nettles. If you killed the nettles, there would be no caterpillars left to eat them so that the nettles would thrive.

We had orchids and bog bean, both beautiful and harmless. We had single patches of wall germander and fox-and-cubs, which the book said were both rare and so they had obviously to be encouraged. Then we had the dreaded ragwort that was poisonous to the cattle. A very useful cop-out was to designate various of the scrubbier areas of the farm as nature conservation areas, where the flora and fauna were allowed to multiply in peace. The saving in hard work allowed by this simple device was tremendous. Over the rest of the farm, the warfare was continuous. I became more sophisticated, latterly, and borrowed a sprayer to attach to the back of the tractor, and with this I would tear over the fields dispensing death.

The one piece of machinery bought that was not absolutely essential for working the farm was a mower. I bought two in fact. The first was a finger-bar mower that worked on the same principle as a pair of barbers' clippers. This stuck out of the side of the tractor and vibrated the grass and weeds to death. It worked beautifully for its first ten minutes of use and then it hit a stone, whereupon it underwent a spectacular disintegration and spread itself in a multitude of tiny pieces over a wide area. I replaced this with a considerably more robust rotary job, still very second-hand, that had the merit of being totally enclosed so that any disintegration would be retained within its own orbit. The machine was mainly used for topping docks and thistles, at which it was adept. However, it was really several sizes too big for my ancient rust-spewing tractor. It used to take half a minute or so before the tractor could rev the mower up to its working pitch and then, with a violent jerk from the clutch, it would lurch into motion. From then, it was unstoppable since the brakes of the tractor were an abstract conception at the best of times and the momentum of the two-ton mower would brook no argument.

When this machine met a stone of any substance, it did not disintegrate but hurled it aside like a cannon ball. If luck was on its side, the stone would pound the interior of its armour-plated casing emitting a puff of rust and a few of the remaining flakes of paint that whizzed about like shrapnel. Otherwise the rock would bound away in front of the machine, galvanising any cow within half a mile or so into instant flight. I used this machine to attack the trees and gorse that formed the impenetrable maternity area in the middle of the bog. Getting it across the bog was quite exciting in itself. I speeded up to 20 mph and ploughed my way over preceded by a bow wave of mud, rushes and tadpoles. Then I found that I could not stop and sailed through the gorse, ending up in the stream.

Normally the gorse was five or six feet high, but the Great Blizzard had settled on it like a great lead bolster and flattened it, leaving a foot of tangled sticks and scrub with a few larger trees poking through. The mower fell on this mess with an enthusiasm that bordered on the maniacal, grinding and crushing it in its maw and spitting bits out into the surrounding bog and woodland. The cows gathered to watch at the edge of the bog and stood in a semi-circle with admiration on their faces.

All the excitement was altogether too much for the tractor. It gave up the ghost by catapulting its radiator cap high into the air followed by a great column of steam that brought a spontaneous moo of applause from the gallery before they took to their heels in alarm. The tractor did not often let me down. It lasted five years for an initial outlay of a hundred pounds and I grew to know and love its little foibles, as did the cows. They used to stand beside it and lovingly lick all the grease and dirt off the engine so that it was always immaculately clean. It was never very strong on fuel pipes and as the metal cracked and disintegrated over the years, they were gradually replaced by yards of rubber milk tubes. It also, as on this occasion, used to consume fan belts.

The machine was prepared to tolerate bits of baler twine and the odd pair of nylons as very temporary replacements but these it could destroy with contemptous ease. The art of fitting a new fan belt correctly was too clever for me. The shaft over which the belt was slipped was supposed to loosen slightly to enable it to go on. This may have been the case several decades earlier when the tractor had been in its infancy, but no longer. All parts that were not forced to be in continual motion had long ago rusted solidly into place, this shaft amongst them. The fan belt invariably would end up with pieces gouged from it by a screwdriver as it was forced into place and upside down, which would give the tractor pause for a week or two before it would eat it and require yet another replacement.

One of the dedicated safety officials from the Ministry came visiting one day and took a very dim view of the tractor. These days, tractors have smartly painted, sound-proof cabs and, for all I know, flushing loos in the back. Ours was the product of a coarser age. A highly uncomfortable seat perched precariously over a set of wheels and over an engine which wreathed the contraption in blue smoke when it was in motion. It was very dangerous, said the Ministry man. It might overturn and squash its driver. This is one of the less attractive pastimes in which tractors tend to indulge. In fact ours was already practised in the art, having been frustrated in its youth in an attempt to pulverise its driver only by a conveniently placed boulder which had been very slightly larger than the driver's head. This had broken the machine's fall and it had hissed and clicked at him until twelve strong men and true had levered its carcass off.

Since the tractor, having so very nearly tasted blood, could never really be trusted again, I decided to fit a safety frame to it. For only a very few hundred pounds, you could purchase a hoop of iron with a detachable top: the theory was that this arched over the driver and prevented his obliteration should the machine overturn. The frame arrived and I and a local expert fitted it. There was a problem. With the frame correctly positioned, the handbrake, what little handbrake there was, could no longer operate. Another of those interesting agricultural dilemmas. Was the tractor safer with a frame and no handbrake or a handbrake and no frame? We consulted the supplier who enthusiastically endorsed the frame, so we left it in place.

The next little difficulty was that the frame reared up way above the tractor and was higher than the entrance to half the sheds. I only noticed this when I went into the cubicle shed to scrape out and half the door frame came in with me. I soured still further to the frame when I tried to remove its easily detachable top and found that the whole thing was slightly out of true and required twenty minutes' work with a sledge-hammer to decapitate it so that I could muck out the other half. The frame even managed to break the shaft of the hammer when I was trying to replace its top by cunningly dodging the head.

The ultimate betrayal by the frame was during a minute or two of total chaos in the collecting yard. It brought down the gutter while going in, but that was to be expected. The technique used in mucking out the yard involved a considerable amount of ricocheting by the scraper off walls and pillars so that some of the more inaccessible corners of the yard could be reached. The first time I did this, my head joined in by bouncing off the frame. For some time, I thought that I was dead until the sound of trickling blood slowly revived me. Just as I was slowly returning to normal, the tractor's engine suddenly went totally berserk and threatened to blow up. I came up out of the seat like a rocketing pheasant in my well-practised escape and once again my head met the frame with a crunch and I ended up on the floor of the shed in a couple of feet of dung.

It was not the engine blowing up, of course, but the RAF doing their low-flying exercises. One of their jets had stolen up on me at about fifty feet and its approach had been drowned by the noise of the tractor in the enclosed building until it let its after burners rip directly overhead. It seemed as if the world was coming to an end. I ripped off the frame and hurled it into the deepest and most noisome bog on the farm.

There was one time during the year when some quite serious work actually took place on the farm and that was during silage making. The success of silaging had a critical effect on the profitability of the rest of the year. Traditionally, haymaking was the method whereby grass could be stored and fed to keep the cattle alive during the winter, but hay suffers from certain inherent disadvantages. You need a minimum of four or five consecutive fine days in which to make it. That might be no problem in the Sahara but they did not come all that regularly half way up a mountain in the middle of Devon. Hay that has not had the weather to cure it properly is lousy hay and, in certain conditions, can lead to the growth of certain fungi which are poisonous to cattle.

Another drawback of hay is that it has to be cut at a fairly mature stage of the grass's growth. The stemmier and more mature the grass when it is cut, the less feed value it contains. Really well-made young hay is just as good if not better than its equivalent in silage, although both are a rare and exceedingly expensive commodity. One of the horrors of hay, so far as I was concerned, was that it involved hours of sweated labour, humping great heavy bales of the stuff on to trailers and off from trailers into stuffy barns, usually in the heat of high summer with scratchy bits of it getting down your shirt and into your pants. Silage has the virtue of consistency. You can always guarantee to make quite adequate silage whatever the conditions. You can almost always guarantee to make very good silage if the weather gives you any sort of a break at all. A good example of the financial importance of the fortnight during which the silage was made was between the best silage that we managed to make and the worst. To produce the desired quantity of milk from the cows over the winter, winter being the period when the bulk of our milk was produced, with bad silage I needed to feed £5,500 more of bought-in cake than I did with the best; and the worst was probably that much better again than average hay.

To make silage, I carefully examined the pages of the Book and followed its instructions to the letter. We had the good fortune to have a Ministry of Agriculture experimental farm a few miles up the road that specialised in making silage and testing the various methods of its production. Whatever they did, I religiously copied. They said use an additive when conditions, usually wet ones, called for it. They even told me which brand to use. Additives, particularly the ones that work, are thoroughly nasty corrosive acids and other chemicals which play absolute hell with both man and machine, but whose use ensure the correct sort of fermentation within the silage pit. Silage, for those not agriculturally educated, is grass or other green material that is pickled in its own juices to preserve it rather than hay, which is the same material dried.

The first time that I found out how I was supposed to make the stuff, I checked with the contractor to ensure that he would have no reservations about using additives should the conditions call for it. He seemed quite happy. He stayed quite happy until a week before we were due to start when he suddenly decided that he had doubts. In a flood of indignation, I rang round the neighbours with the ultimate result that they bought lots of expensive equipment and contracted both it and themselves out to me when I wanted to make silage. It was a system that worked beautifully.

The annual process would start on the 14th of May when I would go out to the silage half of the farm and part the docks to check on the growth of the grass since the fertiliser had been spread a month or so earlier. Hopefully it would be adequate and I would alert the neighbours to prepare to begin the process within a few days. I would anxiously start watching the weather forecast after lunch on Sundays. As soon as a couple of days without hurricane or typhoon was promised, the wagons would roll. I would scour out the interior of the three silage pits and erect eight-foot lengths of plywood to provide a wall at the end, which would be removed to let the cows in to eat the stuff come winter. Neighbour 1 would turn up with his tractor and mower and, usually on a Saturday, would gallop through the thirty-two acres and leave it to steam quietly in neat rows.

On Sunday, Neighbour 2 would turn up and scatter the grass around a bit to help it dry before lining it up in more neat rows. I would line the interior of the pit with acres of black polythene. On Monday morning the forage harvester and a couple of trailers would come swaying down the lane. The buckrake would be dropped off and I would fix it to my own tractor. My particular

149

corner of the operation was to use the buckrake to actually fill the pit.

The harvester would start lumbering up and down the field at a snail's pace, picking up the grass and hurling it into the trailer behind after it had roared through the chopping mechanism that sliced it into three-inch lengths which made for a better fermentation and for easier eating by the cows. A constant shuttle of trailers would come lurching down the field and drop the grass in huge heaps in front of the pit for me to pick up and place where it should go. The only break in the stream of arriving grass would be when the chopper jammed up with a lump of grass or, God forbid, picked up a lump of metal that had not been rolled into the soil. Then the roar of the machine would die and be replaced by a stream of curses echoing down from the field which, being on the opposite side of the valley, acted like an enormous sounding box.

This would give me a chance to catch up on the backlog of trailer loads that would soon mount up again once the chopper had got back underway. Buckraking was tedious, even though important. I suspect that that was why I had landed up with that particular job rather than ferrying the grass around or whatever. You were no longer interested in driving forwards but in going backwards so that you could carefully position the quarter-ton rake-load of grass in the right place. Initially the grass would be dumped in the back-end of the silo, the floor of which was three feet below ground level, and this drop would have to be carefully filled with grass so that the tractor could reverse on to it and continue filling it towards the front which was fifty feet away.

To an agricultural virgin like myself, it was not as easy as it should have been. For a start, I have a natural antipathy towards machinery of all kinds and having to concentrate on the gear shift from forward to reverse, the lift and lower mechanism on the buckrake and the buckrake tip all at the same time, meant that I would frequently back in to a wall or a fence while desperately wondering if I had remembered to press all the right levers. I would back the tractor into a mound of grass and try to lift it. The front wheels of the tractor would come off the ground. I would shake a little of the grass off and try again. This time the rake would come up with a rush and hurl hundredweights of grass into the air which would fall on my head. I would take it rather more cannily and end up with a lawnmower boxful.

The grass also needed to be very well consolidated before the tractor could dare to venture on to it. Frequently and ignominiously, I would sink up to the axle and would have to wait to be towed out by the scornful teenager who transported the trailers. The pit would slowly fill. I would get to the other end with a three-foot depth of grass and start to build another layer on top. The polythene sheet that was hung from the wall to line the pit would begin to show scars and tears where I had bashed against the side with the prongs of the buckrake, and my eyes would probably be watering due to the fumes of the formic acid additive.

I would need my first of many clean pairs of pants when my carefully prepared bulwark of plywood at the far end would suddenly shift with a crack of timber and threaten to catapult me and the tractor to the ground below. I would get off the tractor and run round to prop it with extra pieces of wood or metal or anything that was at hand to prevent its total collapse, which would spill tons of grass into the yard. The trailers would keep piling the grass up at the other end. Everything would slow down for the evening milking as members of the crew disappeared to look to their cows. Then back again and into top gear for the final run through to darkness. By dark, the roar of the tractor had been bouncing off the tin roof into your head for nigh on twelve hours; your neck would be stiff from constantly craning over your shoulder as the tractor drove backwards; dropping the loads of grass would become more and more hairy as you perched on top of the tons and tons of grass that would be filling the silo and looked down on to the unyielding concrete 20 feet below. The changes of pants would need to be more frequent as you felt the grass slip and slither beneath the wheels of the tractor and lurch it towards the silo edge.

Then the operation would close down for the night. We would run a sheet of plastic over the results of the day's work and anchor it down against any gusts of wind that might disturb the tranquillity of the summer night. The air would hold a tingling silence after the roar of the tractors had died away, broken only by the squeak of a bat hunting for midges and the regular crunching of the grazing cattle as they browsed at the grass beside the shed.

Next morning, I would milk the cows early and, before the others arrived to continue the operation, would strip off the layer of polythene and roll the silage. The principle behind this exercise was to ensure that the grass was as firmly settled as possible with no pockets of air left within the silo. Air leads to the grass breaking down in reaction with the oxygen and one of the by-products of this is heat. The grass could have become distinctly warm overnight, but the covering sheet of plastic prevented any fresh air reaching in and kept the heating to acceptable levels. It has been known for a pit of silage to burst into flames.

Around 10.30 am with the dew dried on the grass and their chores completed on their own farms, the team would arrive to continue the silaging. The machinery would be greased and the chopping blades on the harvester sharpened. The noises would once again echo over the farm. The harvester would now be coming to the end of the twenty-acre field and, no longer able to thunder down its 400-yard length, it would now be hunting around in the shorter rows of grass at the field corners; the slower delivery of the grass would give me more time to spread it evenly over the pit. The trailers turned in a corner of the field adjacent to the pit, and the cows would come down to watch and snatch any freshly cut grass that came within range. Cows are like sheep, having the profound belief that the grass is always greener over the electric fence and they would squabble amongst themselves at the acid-soaked droppings from the trailers, while the grass underhoof would be shimmering in the breeze.

Towards the end of the second day, the first cut of silage (breakages permitting) should be completed. The tractor would be showing some signs of strain as it was not used to a couple of days of continuous hard work — nor was I. Its clutch would be slipping and it would be chewing up fan belts with increasing frequency to give itself a rest. The first pit would be full to the rafters; often the exhaust pipe on the tractor would need to be removed in order to get beneath them. Again the pit would be covered for the night.

The following day would be one of comparative leisure after the rush to keep up with the incoming loads of grass. There would be a final, careful roll of the grass, the tractor crawling across it at ¼ mph. If the pit had not been filled as evenly as it should have been, the machine would dip and sway as it moved over the undulations. The edges would be the most crucial part to roll because in them pockets of air could often form which would cause the grass to heat and rot. In rolling them, a balance had to be struck between bringing the tractor as close in to the edge as possible and rolling over the edge or getting the tractor stuck and ripping up the polythene in the struggles to get clear.

The first time I made silage, I was told that, such was the quality of the grass, it would very likely burst into flames unless it was thoroughly rolled. I laboriously hauled the three-ton ballast roller to and fro over the top of the pit until it sank and became irremovably stuck. It stayed there until the cows ate their way through to it the following winter.

After the pit had been rolled to its own satisfaction, the final seal could be made. The side sheets were spread in towards the centre and the top sheet pulled over and carefully tucked into the edges to ensure an air-tight seal. The whole pit would then be covered in bales of straw to ensure that the sheet stayed in contact with the grass. Done carefully, and when the pit was opened six months later, it would have a pleasant fruity smell. Done badly, it would smell like a dead elephant.

The above description was perfection. It called for weather that never went wrong, with machinery that never broke down and all the workers turning up when they were expected. If all these conditions were fulfilled and drought had not hit the grass yield, then we could be sure of a profitable year ahead. The weather did quite often go wrong. One of the effects of rain was that the cost of the operation immediately went up. The grass-carrying trailers could hold four tons. When it was four tons of dryish grass, fine. When it was two tons of grass and two tons of rain, then the time needed to clear the field could double.

If rain fell on partially wilted grass, further problems emerged. We cut the grass and let it dry in the sun to reduce its moisture content. In perfect conditions, this grass would rattle through the chopper and even trail a fine haze of grass dust behind. Let a little rain fall on the drying grass and it would turn sticky, great lumps would clog the harvester and bring it to a grinding and expensive halt. The neighbours were paid by the hour.

The worst effect of rain was that it softened the ground. Our land was heavy and sticky and it could be disastrous when the tractors broke through the surface and churned their way into mud. Machines became bogged down, trailers overturned, and the grass that arrived at the pit would be liberally laced with mud which produced a sour and unpalatable silage. Even worse, the

Charles Gore

mess that could be left would severely reduce the yield of grass in subsequent cuts of silage.

Breakdowns, both human and mechanical, were usually my own fault. I would bend down to pick up a roll of polythene, give it a heave and be locked in position for a week. My tractor was by far the oldest and tattiest that participated in the silaging operation. Once I was bouncing happily off the grass in the pit towards a newly deposited load when the tractor suddenly buried its nose in the silage and hurled me over its bonnet, while its front wheels shot off into the field. Even I could tell that this was, mechanically, abnormal and all work had to stop while the machine's corpse was removed from the pit and a substitute found.

The substitute was someone else's rather smart tractor which I was not trusted to operate; instead I was put on the job of ferrying the trailers to and from the field. Even there I made a balls-up. I brought down my first load to the pit and started to tip it. Unfortunately, I had neglected to ensure that the towing hook was fully locked on to the trailer and the entire trailer tipped. The tow bar shot up, smashed the back window of the tractor, took my hat off and came to rest on the roof after jerking the back wheels of the tractor off the ground. 'Tut, tut' would be a highly censored version of what came out of its owner's mouth. After giving me an amazing bollocking, he decided to trust me enough to let me continue on transport, believing that I would never dare to err again.

I would attempt to reduce the astronomic silage bill by working with the gang on their own operation, and with my own tatty tractor I would wind down the narrow lanes from the field to his barn with a trailer-load of grass in tow.

During one of these trips, I came upon a tourist coming in the opposite direction. Tourists were a fairly common feature of the local landscape, although they were more often found on the motorways and the beaches than in our rural backwater. This particular tourist was a very typical sample of the breed. He had the usual shiny car with a London number plate. The male was wearing a bright red shirt stretched tight over his swelling stomach; the wife invisible behind sunglasses and two indeterminedly sexed children picking their noses in the back seat. I stamped on the brakes and, miraculously, the wheels locked and I came to a juddering, swaying halt. We stopped and looked at each other for a bit. There was no possibility, in a typical Devon lane, of passing each other. He would have to back.

The locals in this part of the country, through constant practice, can go just as fast backwards as they can forwards. I had forgotten that in more beroaded parts of the land, the same skills are not perfected. The driver, his wife and kids all looked carefully through the back window of their car and it reversed firmly into the hedge.

Fascinated, I switched the tractor off so that I could better appreciate the diversion. He came determinedly out of the hedge and they all looked over their shoulders again. In a series of galvanic jerks, the car leapt over the lane and buried its back-end in the opposite bank accompanied by a crunch and a tinkle of glass. I could see the wife's lips moving in a vicious fashion and he turned a shade of red that beautifully matched that of his shirt. He turned the steering wheel and the car rocketed out of the bank straight into the original hole. I hugged myself in delight and tried to retain that baffled expression which is expected of us country bumpkins. There was much furious discussion inside the car and longing glances were cast at the passing place twenty-five yards further back. This could well have been the best moment I had had since I started farming. Eventually the driver got out of his car and strode towards me. I thought for a second that he was going to ask me to reverse his car for him but that would have been more than his manhood was worth.

'Would you please move your bloody lorry out of the way.' Not put quite as well as it could have been but, under the circumstances I thought it quite a reasonable effort. For a moment I contemplated playing the uncomprehending country idiot to see if I could extract any further comedy out of the situation but he was rather bigger than me and very cross, so I meekly reversed the tractor and trailer back to the nearest gateway and watched them sweep by me; husband and wife gazing straight ahead, one child raising two fingers and the other, I think, giving me a grin.

That was not our only encounter with tourists. We decided to exploit them by putting a caravan in the corner of a field and stuffing it with as many of them as possible. They were a curious and assorted bunch. We put an electric fence round them to keep them away from the cows but they did tend to mingle. The cows seemed to get more out of it than the tourists and would spend many hours a day leaning over the wire and staring at them, to the acute embarrassment of a couple of the stouter visitors who tried to get a suntan in skimpy bikinis. An entire family moved in and stayed during a week when Joe was at his busiest and they watched his Lothario-like progress round the herd with amazement. The son kept a tally on how many times Joe managed it. The father was very grateful to me as he told me that it was the best course in sex education for his children that could be devised.

Another family were encouraged by the cows to take a regrettable interest in all aspects of the farm which meant that I had to spend large portions of the day driving them away from the slurry pit and stopping them from smoking in the straw stacks and, worst of all, having to put up with a string of loud and inane questions while trying to milk. My cup ran over when one of them asked a particularly loud and inane question just below and out of vision of one of the cows who nervously voided her bowels and bespattered the brute from head to toe. The family got its revenge by using a couple of towels to clean down the beloved.

Another couple reported with great excitement that two golden eagles had been having a fight on the roof of the caravan. No, it had not been a buzzard flying past but definitely golden eagles, two of them. They knew because eagles used to visit their bird table in Croydon occasionally.

The husband of that particular duo was taking his hair-shedding white dog for a walk nearby when, with a flourish of horns, a pack of hounds burst from a nearby hedgerow followed by a score of horsemen and came up the lane towards him. He turned and fled as fast as his heels could carry him, convinced that he had fallen victim to the quaint Devon custom of hunting the tourist.

Apart from the mind-boggling job of emptying the chemical loo between visits, the tourists provided an easy income. It was Joe who killed the golden goose by ripping down a fence during one of his red-blooded periods and indecently assaulting somebody else's heifers. The owner of the heifers was a member of the local planning committee before which promptly arose the subject of illegal and unpermitted caravans and the crackdown thereupon. Ours submerged under a flood of irate local government officials.

# Chapter Ten

Rabbits, in moderation, are not too much of a problem on the average dairy farm. After all, our farm only grew grass and there always seemed to be a fair covering of that about. The foxes and badgers would take a few, the buzzards would take one or two more and there was always the chance of one for the pot when I could summon up sufficient bloodthirstiness to shoot them.

One spring, something seemed to have upset the equation. There seemed to be rabbits everywhere and I was beginning to intercept outraged looks from the cows at being asked to share every mouthful of grass with half a dozen rivals. We tried to find a solution within the resources of the farm; night-time torch-lit shoots had some success. Slightly better was the careful stalk shielded by a grazing cow. I would leap from behind her at the last moment and pounce upon the rabbit. The snag was that I ended up holding a small trembling furry object which was usually brought back to the house to be treated like a pet by the children which rather negated the point of the exercise. The only solution appeared to be to pray for myxomatosis or call in an expert.

Prayer did not bring the required results, so I made soundings in the neighbourhood and it emerged that the greatest rabbit catcher for miles around was Harry. Gone were the days when the the parish could support two full-time rabbit catchers whose quarry would be sent up to Smithfield on the London train each day. These tyros had now retired, but Harry was the nearest thing to them still operating.

Harry turned up on a Saturday morning. Just as the City businessman needs a black jacket and bowler, so Harry needed his great boots, ancient tweed jacket and mouldering corduroys. He came complete with a twitching sackful of ferrets, a curious whippet-like dog and a young assistant bearing an elderly twin-hammered shotgun.

I introduced myself and Harry fixed me with a piercing gaze. 'When the wind of March be in the East, we'll easy catch the little beast.' He looked carefully at me to ensure that I had caught the full flavour and authenticity of this wise country saw. Harry's coversation tended to remain in this rather opaque and elliptical vein all morning, which made communication with him rather difficult. I would ask a question and be overwhelmed by the accumulated wisdom of five hundred years of rabbit catchers, very little of which could I understand. His acolyte, Boy, had not yet mastered this knack and most of the information on what Harry was doing came from him.

I took them to the most rabbit-infested hedgerow and stood clear to allow the experts to move in. Harry sucked his lips, peered down holes and had a long discussion with his dog. Eventually a strategy was decided upon and he draped his net over the burrows, stationed his artillery on the far side of the hedge and thrust his arm into the sack. He hauled out a stinking white ferret.

'This be middle ferret,' he informed me. 'Big ferret don't like beech hedges and 'tis too early for little ferret.' I forebore to remark that the hedge was a mixture of everything but beech and stood back to await developments. He fed middle ferret down the one hole that he had left uncovered and the dog began to bark hysterically. The ferret appeared at the mouth of the next-door burrow and sat down to scratch. Harry stared at it dully and threw a stone in its direction. A rabbit showed itself at the mouth of one of the burrows, stepped round the net and took off across the field. I pointed this out to Harry who stamped on the ground and shouted 'Get'm, Beauty.' Beauty continued to caper round our feet and bark. The rabbit disappeared over the horizon.

A sudden massive explosion from the opposite side of the hedge alerted us to the fact that things were beginning to happen over there. This was the first of twelve cartridges expended, the only achievement of which was to add significantly to the lead content of our pastures. We subsided back into silence and watched the nets. After ten minutes, Harry muttered 'The rabbits be deep.' After a further five, he moved over to the holes and put his ear to one and listened. Meanwhile big ferret and little ferret had escaped from their sack and were wandering about, peering shortsightedly around them. Middle ferret came out of his hole and joined them. Harry continued to listen, pausing only to swear at Beauty who was still jumping up and down barking.

'Here, Boy, bring the spade round. The ferret be stuck.' Boy stepped over the hedge, waving his cocked gun in our general direction. Beauty and I stepped behind Harry.

'You dig there, Boy. The ferret be an arm long stuck.' Boy looked at the three ferrets who were still meandering about by their sack.

'How many ferrets did you bring out, Harry?' Harry swept Boy with a contemptous look. Ferreting Boys should be seen and not heard. Harry opened his mouth to demolish him when he spotted his three ferrets. He said instead, 'Gather up the nets and we'll move over there.'

The morning wore on. Rabbits came out into the field to eat and play. The ferrets popped in and out of holes and scuttled up and down the hedgerows. The odd rabbit would emerge from the odd hole that Harry had omitted to block, and Beauty and I would watch them as they ran across the field to join their fellows. Harry remained utterly phlegmatic and Boy continued to hang on his every word.

The only time that I opened my mouth was to point out a large and well-used hole that Harry appeared not to have noticed, but the words died on my lips when Harry swept me with another of his contemptuous looks. I did not dare gloat when a rabbit emerged. The only bag was a drinking trough, shot by Boy during one of his wilder moments.

Eventually Harry straightened himself and turned to me. 'Sod this. It's World of Sport in half an hour.' He gathered in his nets and ferrets and without a farewell or a backward glance disappeared back into the primeval countryside. The rabbits got myxomatosis the following week.

Not all the visitors to the farm had the fire and passion of Harry. A fairly early one was the local beekeeper who wanted to place some thirty hives down by the stream on the payment of a jar of honey rent per hive per year. This seemed an admirable idea until I pointed out the badgers which, we agreed, would probably polish off the lot. The venue was changed to a piece of waste ground up by the road, feet deep in bracken and tree stumps which the apiarist was forced to clear before he could set up his hives. For a couple of seasons, all went well. Tractor drivers tended to tread delicately when working in the adjacent field as, on hot days, the roar of the bees could even overwhelm the sounds of the tractor engine. Then outrage. Someone stole a couple of combs of honey and smashed a hive. The beekeeper, after an accusing glare or two in my direction, repaired the damage and all went well until another violation a couple of months later.

The next time, we were ready for him. The beekeeper had over two thousand hives scattered throughout the county. From these, through careful selection, he gathered five hives that contained the most foul-tempered and cantankerous bees that he could find. These were carefully set up on the outside perimeter of the rest of the hives, each delicately balanced and tied to its neighbour by a piece of string so that if one should be disturbed, so should they all. They won a famous victory. A week later, the five hives were askew. There was a trail of discarded tools and corpses of bees; embedded in the gate post of the field was about a hundred pounds'-worth of pieces of motor car. The thief never returned. I kept the tools.

'There are two wasps in the dining room,' announced my wife. They were not wasps but a couple of refugees from the hive city and they were gently ushered out of the window before before the guests arrived for tea.

'Look at those wasps,' said the guests. 'There are hundreds of them.'

'They are not wasps, they are bees. And there are only five.' They were let out and we all went for a nice country walk.

On our return, an ominous thunder filled the house. It was the sound of Heinkels over London in every B war film ever made. This time there were hundreds of 'wasps'. One honeybee in the house, rather like one spider in the bath, is something to be rescued and cherished. Hundreds of the things may be all right in the hive but in the house, rather like a bathful of spiders, they are something that causes the milk of human kindness to curdle and a murderous fury to take over. The bees were filtering down into a disused fireplace and investigation showed a shimmering cloud of them round the chimney pot. A swarm appeared to be casing the joint with a view to moving in. We decided to smoke them out.

The chimney proved to be too disused to allow this and the house filled with wreaths of foul-smelling smoke and some

161

hundreds of thoroughly disgruntled, coughing bees. Another assault was spearheaded by several aerosol cans of insecticide which made them even more annoyed and we retreated in some disarray, leaving the bees in triumphant occupation of the house. It was time for the expert.

The bee king festooned himself in veils and gauntlets, climbed up to the chimney pot and examined the fireplace. He found it most interesting. In his opinion, there had been a colony peacefully living inside the chimney for years and it had been invaded by a fresh swarm. Those coming down the chimney were the casualties and those lacking the necessary moral fibre for the conflict raging above. We had two alternatives: we could either demolish the gable end of the house to clear out the chimney, or we could wait until the situation stabilised which should happen in a few days with the total victory of one or other side. The situation never did stabilise. We blocked off the chimney with a sheet of hardboard and sealed it but we still had bees filtering through, and occasionally the weight of the dead would bring the board crashing into the room. Between times it was really quite peaceful, three rooms of the house close to the chimney breast were continually filled with the drowsy humming of millions of bees behind the brickwork.

Spring was very much the swarming time for another species, the Reps. They came in all sizes and degrees of usefulness, ranging from the smart variety, graced by the title of adviser, down to the bloke who was quite liable to chuck the odd calf or tool into the back of his van on the way out if you did not give him an order. The majority of reps had two things going for them: the rules of common courtesy and the realisation by their victims that they were only doing their job. Against them was the fact that agriculture must be the most over-repped industry in the country.

For every farmer, there appear to be at least two reps and the bulk of them came visiting just when I was about to do something constructive. Two of the reps were always welcome. One was very high powered and knew far more about all forms of agriculture than I did. He had, in fact, been a highly successful dairy farm manager for someone else and I fed him coffee for hour after hour while I meticulously picked through his brain for information and advice which was then put into effect.

The other was welcome purely for his entertainment value. He saw himself as being very much in the same sort of mould as the first with a fund of priceless knowledge on all things agricultural. I had been keeping cows for about a year with fairly dismal results when I had been following his advice. It then dawned on me that he had not got a clue what he was talking about and from then on we got on together beautifully and I spent many happy, if unproductive, afternoons extracting more and more outrageous agricultural ideas that were put forward as the gospel truth. It was he who forbade me to touch a plant that grew in a small patch in a field as it was deadly nightshade and would kill me and all the stock unless I called in an eradicator to remove it. He even alarmed me slightly until this buttercup bloomed. We spent an entire morning tramping the land and taking small samples of soil for later analysis back at the house, and when he had all his apparatus rigged up, he found that he had left his instruction book behind and could not tell me what the results meant.

The arrival of one rep used to send me fleeing in terror to the thickest patch of undergrowth on the farm where I would hide for half an hour until he had gone. He would tour the farm, calling plaintively until eventually giving up. His trouble was that he was too well trained and thorough, quite apart from his product being totally useless. He would plough grimly through his sales manual, asking me leading questions to which he would demand a reply before continuing. At the end of his lecture, he would practise his closes of sale, running the gamut from A to Z until I was a wreck, and would sometimes be forced to place an order just to get rid of him. I cancelled these orders after he had gone.

The desperate lengths to which some people are driven to get rid of or avoid reps can lead to problems. I know of one farmer who came up to the house and nearly fell into the lap of a particularly dreaded rep. He took the only course open to him and nipped into the boot of his wife's car to wait until the coast was clear. Unfortunately he mixed up his cars and hid in the rep's boot by mistake. He sneaked out at the next call ten miles down the road and had to hitch a lift home.

Things always become really hot with the advent of the cake rep. Cattle cake, to the farmer, is a preoccupation second only to slurry. It is usually by far the largest cheque that he lays out each month and the compounders subject him to enormous pressures to ensure that this cheque goes in their direction. Cake creates a fundamental dichotomy in the business of dairy farming. The farmer is interested in the margin between his costs and the amount of money he receives from his sales. Painfully obvious perhaps, but he can take one of two totally different paths to achieve the best margin. The classic path taken by the British dairy farmer has always been to squeeze as much milk as possible out off his cows and to do this he feeds lots of cake and makes the miller very happy.

The alternative is to accept a much lower yield from the cows and cut the cake bill right back. At the beginning of 1980, a high-yielding cow could give 7,000 litres worth £840 but might eat cake to the value of £300, leaving a margin of £540. The same margin could be achieved by a farmer whose cows produced only 5,000 litres of milk but merely ate half a ton of cake. Which of these two paths the farmer decides to take is crucial and obviously affects the running of his whole business. If he tries to compromise, he can often end up with the worst of both worlds.

The cake firms battle amongst themselves for the farmers' business and also battle to ensure that as many farmers as possible go for higher yields which gives the cake firms the greater profits. On the other side lie the fertiliser firms who also battle amongst themselves for business and, since following Path 2 leads to greater reliance on the farm's grass and therefore greater sales of fertiliser, they also tend to pitch against the cake firms.

In the middle lies the farmer who is inundated by propaganda from both sides. Choosing the more profitable of the two paths is no easy decision to make. The prices of feed, fertiliser and milk vary the whole time and what might be profitable on paper for one half of the year can equally be the less profitable for the other half. My own leanings were always towards lower yields and lower cake bills, not for any particularly sound business reason but because I distrust large, powerful organisations that try to make a profit out of me, and to utilise the services of the cake firms in a big way, would call for an even larger overdraft. Since we were always reading about surpluses of milk, it also seemed a bit stupid to go on adding to them if there was a perfectly viable alternative.

This declared policy did not prevent some fairly powerful lobbying by the compounders. One said that he would guarantee that I would make more money the following year than in the previous if I used his cake, a prospect that seemed quite attractive until I became lost in the maze of small print that hedged the offer and rendered it totally incomprehensible. The criterion I used in buying cake was to choose the cheapest supplier of the grade that I required provided he was willing to declare what went into it. It is astonishing to realise that the biggest compounders refuse to tell the buyer what they are buying. You can pay them £150 a ton for an anonymous-looking grey nut and you are not allowed to know its ingredients.

Compounders make errors. You can receive the wrong kind of cake or you can receive a rotten load, and getting a refund from them is like trying to pull a tooth from a shark. One neighbour had a rotten load delivered and the rep came along to negotiate on behalf of his company. After reaching a compromise, they went to examine the offending cake which was in one of those tower hoppers. The rep climbed up the ladder in his smart suit and looked in. 'It's not too bad,' he said. At which the farmer took the ladder away and chucked the rep a shovel and brush and said that he would return in three hours to let the rep down provided he cleaned out the hopper.

The power of the meal rep over me came from two directions. He could take and analyse samples of silage and he could work out cow rations. When I started cow farming, rations were calculated in starch and protein equivalents and calories. Just as the wonderful glow of understanding spread through my brain and all became clear, the entire system was scrapped, the calorie became redundant and I was back to square one with a mysterious thing called metabolisable energy. Reps were the only people I came across who could understand the system, having had it drip-fed into their skulls at innumerable sales training meetings. Some of the brighter reps could carry me along with them in their calculations and, for brief moments, I could share with them the joys of comprehension before I would slip back into the slough of ignorance. It really would have been useful if I had had the system taped. The cake men use to prove to me each year, just how much I needed to buy their product, while the fertiliser man would turn up the next week and prove that I did not. However much I worried and agonised over it, the cows just churned their way through whatever was on offer and seemed content.

Another brand of visitor were those who carried black briefcases marked E11R. The first of these that I met turned up in a mini wearing dark glasses and speaking rather like Humphrey Bogart. I met him down in the yard.

'Hullo,' I said, as he got out of his car. He looked shiftily round him. 'Can I help you?'

'I hope so,' he replied enigmatically, speaking out of the corner of his mouth. 'I'm from the Dept. of Social Security. Here's my I.D.' He pulled out a piece of card from his pocket containing a photograph of a mild-looking individual at total variance with the mac and dark glasses standing opposite. 'You're Mr Robertson, and you're not on Social Security.'

'Absolutely right, so far,' I replied.

'We're hoping that you might help us.'

'In what way?'

'We're looking for information on any of your neighbours that might be working and claiming sickness benefit.' This was a remarkable request.

'I'm very sorry, but I've only just arrived on the farm and the nearest neighbour is a mile away and I don't even know who he is yet.' He slid back into his mini and powered back up the lane. I thought he must be the front man for a burglary syndicate, so I phoned up the local office to check whether my caller had been genuine. The girl at the other end burst into peals of laughter. 'Ah, you must have had a visit from our Inspector Clouseau.' The gallant inspector was obviously hot on the trail of some malefactor because for a couple of weeks I kept coming across him lurking in hedgerows with binoculars. I would receive a grave salute from him. I quite missed him when he disappeared.

The vatman was another regular. In spite of the bank manager's efforts, high finance and myself never really understood each other. The vatman's first visit turned out to be an unqualified success. He had spent the morning going through my books and had built up an ominous pile of notes and was muttering darkly about calling in the enforcement branch when the news came through that Harold Wilson had resigned. He immediately showed the true impartiality of the Civil Service by bursting forth into noisy celebration. I was able to divert his attention to the nearest pub and he corkscrewed off at closing time, leaving all his incriminating notes behind which were hurriedly burnt. I learned eventually how to handle them. They would arrive and go through my books and come up with some figure that they thought I owed them. I would make out a cheque for the amount and claim it straight back again the following month. That way everybody was happy.

The most powerful visitor to the farm was also the most regular. He was the tanker driver. As well as merely picking up the milk, he also read the dipstick and therefore made the decision on how much milk we were actually paid for. Reading the dipstick may not appear to give much room for manoeuvre but a fraction of an inch could made the difference of £5. Multiply that by 365 days of the year and it begins to mount up. If the driver did not like you, he could use his bare finger rather than a tissue to rub the dipstick down before taking the reading. This meant that the milk would register a vital point or two lower down. It was a constant, friendly rivalry. You hoped it was friendly anyway. As the driver came down the lane, you would give the milk a quick stir with the paddle. This aerated it and again led to a higher reading.

Sometimes things went beyond the accepted conventions of the game. I had a visit from the dairy's quality control chief and he opened my eyes to some of the fiddles that went on between the farmers and the drivers. A packet of cigarettes on offer every morning so that the driver's eye would err on the right side, or straight cash bribery. He made out that dairy farmers were a pretty dishonest lot. One that had come his way kept eighty Friesians and one Jersey. When the milk was to be sampled for quality, a bucket of the Jersey's milk was always on offer and that was the milk sampled as representative of the herd. That cost the farmer quite a lot. Another farmer used to add fifty gallons of water to his tank before being expensively caught out. The neatest trick was another who bought milk powder, highly subsidised by the EEC, and placed it in the bottom of his tank and just added water. He managed to keep selling milk for six months after he had sold all his cows. Then he was caught out after he had failed to do a sufficiently thorough mix and gritty bits of powder were found in a quality sample. My corruption was restrained to a bottle of whisky at Christmas and this seemed to be within bounds.

# Chapter Eleven

A farm sale is always rather sad. Ours was no exception. The farm itself was sold in the autumn of 1979 and our cows went that October. My landlord and father-in-law wanted to sell the farm and I felt that I had taken the self-management principle about as far as it could go in that particular situation. The farm ran like clockwork — very profitable clockwork — but there was really no great future for it. Granted, it would continue to earn a living over the years but it would have been stagnating unless I had pumped in vast sums of money to alter all the buildings and the system, and that I was not prepared to do. We decided to sell up and start again somewhere else. With a bit of luck, we would not lose financially as the market was just about to enter a downturn and our landlord was very generous with our redundancy pay; so we should be able to buy our way back in before the market picked up again. I was beginning to feel that I knew rather more about dairy farming than our cows, and a breathing space while we sorted out the way that we would operate in the future would be very valuable.

We discovered that there were two ways that the cows could be sold. One meant huge sums of money being given to us by the EEC, the other meant that we would have to take our chances on the open market. Choosing the EEC grant would mean that all the cows would have to go to slaughter, although there were vaguely dishonest fiddles that could have been brought into operation to avoid this. However, the big disadvantage was that the farm would not be allowed to hold dairy cows for a minimum of five years and it seemed a bit unreasonable to saddle the farm with a restriction like that when it was just going on the market and dairy farming was about the only way that it could be farmed with any expectation of profit. Our cows would be sold in a dispersal auction on the premises.

There are some things that are really desirable when you have a farm auction. You require healthy cows, good weather and a healthy vendor as there is a lot of running about to do before, during and after the sale. I went down with mumps a few days beforehand. It was a little bitter I thought. The doctor thought that it was a huge joke and made me promise to keep out of the ring because, he said, if I swelled up where he expected me to swell up, the buyers would have some difficulty in deciding which one was me and which was the bull. How I laughed. In the event, it was not too bad an attack, I looked slightly more grotesque than usual and avoided oranges like the plague, but few potential buyers fled from me in panic for fear of infection.

The auctioneer doing the selling was the one who bought for me, and on getting an estimate of the likely size of his commission, I did everything but give him a kiss to try and give him some of my germs. It did not work because it turned out that he had already had them.

Some of the cows decided to go ill on me; one had to be withdrawn from the sale when she got summer mastitis and she was the only one that ever died on me for that cause. Some had bad feet and the vet spent several hours splinting and injecting them and they all pulled through all right.

The morning of the sale dawned bright and clear with the cows in a thoroughly bad mood. Their routine had been severely interrupted as I had milked them horribly early in the morning — in fact, almost the night before — so that their udders would all be nice and big and look as if they were giving lots of milk. That they were. Nothing I seemed to do had any bearing on the quantity that the cows gave from one lactation to the next. But, for some reason, they had all decided to be generous that lactation and they were churning the stuff out in a highly impressive fashion and we could announce some very good figures at the sale.

The auctioneers laid out the ring in the middle of the yard and bale after bale of straw was thrown about to keep the cows clean and cover the dung that layered the concrete. It was all part of the subtle art of salesmanship. The cows were pulled out of their field and numbered and put in the cubicle pens ready for sale, and the buyers turned up in their droves. Then the heavens opened and it poured in the third and final boot-sinking rainfall that we had on the farm. I think the one thing that saved the sale from becoming a disaster was my neighbour over the hedge. He had been watching my progress over the past year or two and the fact that we seemed to be making a reasonable living. This, he was quite sure, could have nothing to do with any skills that I might possess because I did not have any. Therefore, ran the reasoning, it must be because I had a first-class bunch of cows. He wanted to buy at least twenty-five of them, and he and I had been through the catalogue and worked out which would be the best ones for him to have.

If it had not been for him, I would have been virtually the only person apart from the auctioneer in the ring as all the other potential buyers had made a rush for shelter in the sheds. I stood wearing my thornproof tweeds, which I hoped was the sort of thing that a farmer was expected to wear at his own sale, and got thoroughly soaked. The water came up to my ankles and beyond, ruining my gentleman farmer's type suede bootees. Some bright spark then shifted some straw that he thought was blocking a drain and a mighty Niagara of water swept through the yard forcing us to cling grimly on to the rails surrounding the ring to avoid being swept down to the stream in a torrent of half-eaten hot dogs, slurry and empty coffee cups.

I could not take it at one point and went to shelter in the cubicles. Joe was in with a bunch of heifers and in the same pen was a family sitting having a picnic. Joe was licking the sandwiches that they were offering him.

The cows sold quite well — very well considering the conditions. I had managed to persuade my neighbour that all the cows that I really liked were the ones that he ought to buy. He did very well. Fear of drowning made him lose track a bit of what he had actually bought and what he had not, and so he missed some when he thought that he might have gone over his budget but only one that I wanted him to have went to another buyer.

The star of the show was Joe. I was expecting to have to keep him as a pet because I could not see a very strong market for a bull that helped himself to £500 a year in milk and I was damned if I would see him go to slaughter. He went for very nearly double his slaughter value and, I am told, now has a constant supply of nubile young heifers at his disposal. He went in a blaze of glory. A neighbour had put half a dozen heifers that had not yet gone to the bull into the sale and, after he was sold, he managed to get amongst them and create havoc. I was told by the worried auctioneer to do something about it and went down and saw an interested and admiring gallery of spectators, none of whom would dare to get involved. I pulled him off one of his harem and for the one and only time in his life, I dragged him up to the bull pen and shut him in.

There was not an animal left on the farm the following day.

The truly rich man is the man who can say that he enjoys what he does. As a dairy farmer for five years, I was a millionaire. We can grouse because it can be damn hard work, and milking every day of life can become a chore of formidable proportions, but I never met another farmer who wished that he was doing something else.

The joy of the industry for me was summed up early one summer morning just as it was getting light. There was a mist lying over the fields and I was about to go out and chase in the cows. With a certain optimism, I stood at the gate and whistled for them. Out of the mist thundered the entire herd of fifty-five and came milling round me. It was one of those moments that sent a shiver of pleasure up my spine. A job that can give you that is a job from which you do not walk away.